Satyajit Ratha · Aneeya Kumar Samantara

Supercapacitor: Instrumentation, Measurement and Performance Evaluation Techniques

Springer

Satyajit Ratha
School of Basic Sciences
Indian Institute of Technology Bhubaneswar
Bhubaneswar, Odisha, India

Aneeya Kumar Samantara
School of Chemical Sciences
National Institute of Science
Education and Research
Khurda, Odisha, India

ISSN 2192-1091   ISSN 2192-1105   (electronic)
SpringerBriefs in Materials
ISBN 978-981-13-3085-8   ISBN 978-981-13-3086-5   (eBook)
https://doi.org/10.1007/978-981-13-3086-5

Library of Congress Control Number: 2018960204

This Springer imprint is published by the registered company Springer Nature Singapore Pte Ltd.
The registered company address is: 152 Beach Road, #21-01/04 Gateway East, Singapore 189721,
Singapore

Dr. Aneeya Kumar Samantara would like to dedicate this work to,
His parents, Mr. Braja Bandhu Dash,
Mrs. Swarna Chandrika Dash and his
beloved wife Elina

Dr. Satyajit Ratha would like to dedicate this
work to,
His Parents, Mrs. Prabhati Ratha and
Mr. Sanjaya Kumar Ratha

Preface

With the revolutionary advent of supercapacitors as next-generation energy storage devices, a significant number of materials have been characterized as supercapacitor electrodes with few of them showing excellent electrochemical activities. The overall performance of a typical supercapacitor device fabricated using these characterized electrode materials depends upon the constituent components of the device besides few key parameters such as working potential window, equivalent series resistance (ESR), energy density, power density. The range of calculation parameters used to evaluate the performance of a supercapacitor varies, thus creating few significant hurdles for the device to be realized in practice. This book is an attempt to recollect the calculation methods that have been implemented till date and focus on the key factors or parameters including any discrepancy arising out of the instrumentation section affecting the same. This will help not only to determine the appropriate methodology for the evaluation of a supercapacitor device but also to single out the inconsistencies between the device performance and electrode material properties.

Being a promising energy storage option, supercapacitors should be readied for urgent commercialization, rather confining them within the niche of academic research. Therefore, strict performance guidelines aided by realistic evaluation techniques should be followed which would help make these wonderful devices reach out to the society through the industrial measure. In order to execute these objectives, a brief discussion on instrumentation of the supercapacitor, different measurement procedures, and the techniques used for the performance evaluation is presented in this book.

The first part of this book furnishes a brief introduction, different instrumentations, and their measurement. Afterward, a detailed discussion on the techniques used for the performance evaluation of supercapacitor is presented. This book has been written keeping in mind that the broad readership will graduate students, academic researchers, and industries involved in sustainable energy and growth. No other publication has addressed these areas so comprehensively, and therefore, this book can be considered to be highly original in content, with no competing texts.

Bhubaneswar, India   Satyajit Ratha
Khurda, India   Aneeya Kumar Samantara

About This Book

With extensive research on electrical energy storage, there has been a steadfast approach from the scientific community to characterize a broad range of compounds in order to design efficient and high-performance supercapacitor electrodes. Considering the myriad of such compounds, there must be characterization of tools and techniques, which is diverse enough to provide detailed charge storage aspects of such materials. These simple techniques range from cyclic voltammetry to more complex impedance spectroscopy or step potential electrochemical spectroscopy. However, the proper evaluation of material can only be achieved with a clear understanding of the underlying principle behind the charge storage mechanism which could otherwise lead to severe ambiguities. This book clearly emphasizes the required methodologies and technical interpretations so as to identify the charge storage mechanism of a specific material (whether it is EDLC, pseudocapacitive, or faradic). Subsequent discussions provide a set of precise and detailed mathematical expressions to carry out the proper evaluation of a supercapacitor device. The authors have also discussed the inconsistencies that are often encountered during the comparison of data obtained with different measurement fixtures and the necessary steps to correctly address them. The overall aspect of this book is to sensitize researchers (particularly academic) working in the relevant field to make supercapacitors effective through a vivid approach.

Contents

About the Authors

Dr. Satyajit Ratha pursued his Ph.D. at the School of Basic Sciences, Indian Institute of Technology Bhubaneswar, India. Prior to joining IIT Bhubaneswar, he received his Bachelor of Science with first-class honors from the Utkal University in 2008 and Master of Science from the Ravenshaw University in 2010. His research interests include two-dimensional semiconductors, nanostructure synthesis and applications, energy storage devices, and supercapacitors. He has authored and co-authored 20 peer-reviewed articles in international journals and 1 book.

Dr. Aneeya Kumar Samantara is currently working as a postdoctorate fellow at the School of Chemical Sciences, National Institute of Science Education and Research, Khordha, India. He pursued his Ph.D. at CSIR-Institute of Minerals and Materials Technology, Bhubaneswar, India. Before joining the Ph.D. program, he completed his M.Phil. in chemistry from the Utkal University and Master of Science in advanced organic chemistry from the Ravenshaw University. His research interests include the synthesis of metal chalcogenides and graphene composites for energy storage and conversion applications. He has authored articles in 19 peer-reviewed international journals, 1 book, and 3 chapters. Further, two books and three research papers are in press.

Abbreviations

2D	Two dimensions
3D	Three dimensions
CCCD	Constant current charge–discharge
CD	Charge–discharge
CDCs	Carbide-derived carbons
CNF	Carbon nanofiber
CNT	Carbon nanotube
C_T	Total capacitance/net capacitance
CV	Cyclic voltammogram
DC	Direct current
EDL	Electrochemical double layer
EDLCs	Electrochemical double-layer capacitors
EG	Exfoliated graphene
EIS	Electrochemical impedance spectroscopy
EMD	Electrolytic manganese dioxide
ESR	Effective series resistance
GCD	Galvanostatic charge–discharge
GO	Graphene oxide
HCFs	Hollow reduced graphene oxide fibers
LED	Light-emitting diode
m-SCs	Micro-supercapacitors
MWCNT	Multi-wall carbon nanotube
NiOH	Nickel hydroxide
nm	Nanometer
NWs	Nanowires
PANI	Polyaniline
PbO_2	Lead oxide
PCs	Pseudocapacitors
PDMS	Poly (dimethylsiloxane)
PEDOT	Poly (3,4-ethylenedioxythiophene)

PEE	Pulse energy efficiency
PET	Polyethylene terephthalate
PSS	Polystyrene sulfonate
PTFE	Poly (tetra fluoro) ethylene
PVDF	Poly (vinylidene) fluoride
R_{CT}	Charge transfer resistance
rGO	Reduced graphene oxide
RHE	Reversible hydrogen electrode
R_s	Solution resistance
SC	Supercapacitor
SCE	Saturated calomel electrode
SE	Specific energy
SEI	Solid electrolyte interphase
SMSC	Shape memory supercapacitor
SP	Specific power
SPES	Step potential electrochemical spectroscopy
SSC	Symmetrical supercapacitor
s-SCs	Stretchable Supercapacitors
XPS	X-ray photoelectron spectroscopy

Abstract

This book presents a brief discussion on the types of electrodes and their configuration to form flexible supercapacitors, stretchable supercapacitors, and micro-supercapacitors to meet the current requirements to power the wearable and miniaturized electronics. Though different electrochemical measurement techniques are available and various methodologies are followed for the performance evaluations of the supercapacitors, there are still so many inconsistencies associated with them. In this regard, a detailed discussion on the performance evaluation processes is presented elaborately along with a suitable example. The authors presume that this book will help the energy and materials researchers to gather more knowledge in this field and to explore supercapacitors with higher efficiencies for practical application.

Keywords Supercapacitor · Energy storage · Energy density · Power density Capacitance

Chapter 1
Introduction

Being one of the most versatile energy carriers, electricity is predicted to dominate the world economy in the later part of the twenty-first century, rivaling fossil fuels (Green and Staffell 2016). However, little has been done to bring desired improvements in the effective generation and distribution of electricity. The process of implementation of renewable energy resources and subsequent production of electric energy has been overshadowed by the increased use of coal. Therefore, a radical change at the global level is critical for sustainable growth and environment safety. Though existing technologies are capable enough to convert renewable resources into electric energy, their intermittency and lack of necessary infrastructural modifications pose major challenges that have to be taken care of. Effective distribution of electricity is as important as its generation, which is why electrical energy storage plays a pivotal role in creating flexible and reliable grid systems. As the demand for electric energy is expected to increase in the years to come (at an average growth rate of ~3% per annum), a steadfast approach on the global platform to evolve and promote renewable resources (led by electric energy) is the need of the hour (IEA 2017). The most prolific development in the field of electrical energy storage has been realized through battery technology which has turned out to be an integral part of low power grid/backup systems since few decades. Furthermore, the Li-ion technology has brought a significant boost to the consumer electronics, especially portable electronics due to the lightweight nature of Li. Since few years, a significant number of automobile manufacturers have implemented the Li-ion technology to drive hybrid electric vehicles/electric vehicles to curb excessive emission and increase both the efficiency and lifetime. Although these next-generation rechargeable batteries have high energy densities, they lack sufficient power density values. As the crave for power is gradually taking over the energy density, battery technology alone would not be sufficient to provide a solution in the long run.

It is to be noted that the cost-effectiveness of Li-ion technology does not promise much improvements in the coming years considering the limited reserve of Li in the Earth's crust. Supercapacitors, on the other hand, make use of materials that are earth-abundant and non-toxic (fabricated mostly from carbon-based starting materials).

© The Author(s), under exclusive licence to Springer Nature Singapore Pte Ltd. 2018   1
S. Ratha and A. K. Samantara, *Supercapacitor: Instrumentation, Measurement and Performance Evaluation Techniques*, SpringerBriefs in Materials,
https://doi.org/10.1007/978-981-13-3086-5_1

These devices are non-hazardous (not affected by thermal aberrations and can operate within a wide range of temperatures, typically between −40 and 85 °C), can tolerate over-abusing, and have at least 10–100-fold power densities than any battery system which makes them promising alternatives for electrical energy storage. Furthermore, extensive research projects have been taken by the likes of Tesla Inc. (USA), and Mazda (Japan) including core manufacturers such as Maxwell, Nippon, Skeleton technologies, and Yunasko to name a few in order to replace battery technologies with supercapacitors in electric vehicles that would help in curbing the extra weight and improve the safety and stability of the same (Jussi Pikkarainen 2016; Miller et al. n.d.; Schneuwly n.d.; Tomáš Zedníček 2016; "Ultracapacitors Still Showing Promise" 2014; Vaughan 2018). Though present supercapacitor technology is used mostly to provide an assistive storage buffer for the battery stacks in hybrid electric vehicles, the world could see, in coming years, clean, environment-friendly, and efficient electric vehicles being implemented at large scale that would require less (or zero) involvement of the battery technology which would make the charging process much quicker and would significantly improve the life span of the power systems (Kouchachvili et al. 2018).

Electric energy storage via capacitive technology follows electrostatic charge adsorption process and does not involve complex chemical reactions that occur in batteries. This not only enables the supercapacitors to charge/discharge faster (providing high power densities), but also prevents safety hazards (of any kind) making them a cleaner and safer storage alternatives. There is, however, tradeoff between power and energy, determined by the time parameter. High power densities of supercapacitors leave them with modest values of energy densities. Supercapacitors based on materials that are either carbon or its derivatives (such as activated carbon, graphene/reduced graphene oxide) are purely of electrostatic nature; i.e., they rely on Helmholtz double-layer formation through adsorption and coulombic interaction (mostly known as electric double-layer mechanism) for charge accumulation on the electrode surface. Little can be done to achieve higher energy densities in these carbon-based supercapacitors and require extensive material designing to achieve improved energy densities. With rapid commercialization and evolution in technologies, steady efforts have been made to reconfigure the battery systems that would be as fast as supercapacitors, or improve the specific capacity of supercapacitors (Yu and Chen 2016). In this context, materials having excellent reduction–oxidation properties are of great importance, especially those which have striking morphological analogies with that of graphene/reduced graphene oxide (Yu and Chen 2016). Subsequently, the concept of pseudocapacitance has emerged enabling the supercapacitor technology to achieve enhanced energy densities without compromising the power densities. In addition to that, there are a large number of reports on materials that are purely faradic in nature, but have been claimed to be ideal for supercapacitor electrodes (Ghosh et al. 2014; Jiang et al. 2016; Lakshmi et al. 2014; Meher and Rao 2011).

As has been discussed above, supercapacitors are supposed to work in a pretty straightforward method (electrostatic adsorption on the electrode surface) which requires majority of the electrode materials to have large surface area with negligible thickness so as to accommodate electric charge at large scale. Researchers, therefore, resort to a wide range of fabrication methods that would yield thin layers of electrode materials on various current collectors. Preference is given to electrode materials possessing two-dimensional structures or layered structures (with layers stacked via weak Van der Waals force of attraction) which would facilitate the formation of extremely thin layers on those current collectors. The use of bulk materials is mostly avoided to prevent additional processes besides surface kinetics that could slow down the charge–discharge process. Batteries, on the contrary, use bulk electrodes that make the whole process of charging/discharging tricksy enough to impart high energy densities. Though the concept of a device which is part battery and part supercapacitor seems to have drawn significant attention from the researchers, their fabrication method and subsequent investigation of the electrochemical properties require different approaches from those followed in the case of pure capacitive materials. Different approaches should also be made for materials having rapid reversible redox properties (pseudocapacitive). Though part of their electrochemical characterization shows (at least graphically) a behavior similar to carbon-based supercapacitors, there are several techniques which could reveal their true charge storage mechanism.

Since robust storage technology is highly essential, not only in preventing the depletion of fossil fuel reserves, but also to promote the implementation of renewable, utmost care should be taken in the selection of both materials and methods, respectively, in designing and evaluation of high-performance supercapacitor devices. To aid the industrial progress, correct analysis techniques at minute level is required to distinguish between non-faradic (capacitive) and faradic (Nernstian) materials, the absence of which have already triggered few ambiguities at the academic level. There has been a surge in the number of electrode materials, i.e., metal oxides, sulfides, carbides, nitrides, hydroxides, and phosphides including few polymeric substances, which therefore requires critical stratification with regard to their charge storage mechanisms to get further insight into their electrochemical performances (Guan and Wang 2016; Mombeshora and Nyamori 2015; Samantara and Ratha 2018). With new concepts being divulged to amalgamate both the battery and the supercapacitor technologies together to increase the probability of getting a hybrid system which would bring a balance between energy density and power density, researchers (particularly academic) working in the relevant field are in a state of perplexity as to whether there should be a specific set of terminologies and parameters (different altogether from those taken/implemented for EDLCs) to address the electrochemical behavior of electrode materials relying on mechanisms such as intercalation and/or reversible/quasi-reversible reduction/oxidation of central metal atoms to store charge. It should be noted that different electrode materials starting from EDLCs, pseudocapacitors, and faradic would require different sets of evaluation methods (considering the fact that even the charge storage mechanism for EDLC materials like activated carbon or reduced graphene oxide has been misinterpreted by many of us). Therefore, a detailed insight into the evaluation techniques that are currently in practice and their possible modifications is necessary to make the whole process of analysis and interpretation less error-prone and unambiguous.

Chapter 2
Instrumentation and Measurement

The supercapacitor and batteries have different constituent chemistry and follow different mechanisms for their operation. But both of them have nearly similar manufacturing and cell construction processes (Conway 1999). Likewise batteries, the supercapacitor comprises (i) a pair of electrodes containing active electrode materials adhered to a suitable current collector with terminal for connections to the electronic circuit, (ii) a porous membrane that physically and electrically separates both the electrode preventing from electrical shorting, (iii) electrolytes containing the charged ions generally soaked with the membrane separator, and (iv) packaging components that make to place all the above components at their respective positions in a single cell avoiding leakage (Samantara and Ratha 2018). Generally the carbon-based materials like graphene, carbon nanotubes, activated carbon, and their composites with metal oxides, sulfides, selenides, conducting polymers, etc., are considered as the active electrode material. Slurry of these materials in a suitable solvent and binder (occasionally used) are coated at an optimized thickness on a pre-treated metallic current collector to make the electrodes. This coating of the active materials can be carried out on one or both the sides of the current collector depending upon the final cell configuration. Also without using the metal-based current collectors, the free-standing electrodes of SCs observed to be demonstrating good mechanical behavior in form of flexible and stretchable configurations. On the other hand, the separator typically made up of porous polymeric membranes like cellulose nitrate, poly (vinylidene) fluoride (PVDF), poly (tetra fluoro) ethylene (PTFE) membrane. Generally these separator membranes are soaked with the electrolytes (aqueous, non-aqueous, and ionic liquids) and impregnated between these two electrodes. Although, the aqueous electrolyte has better ionic conductivities and is of low cost and convenient for use compared to organic electrolytes, they are associated with a narrow value of stable potential window (thermodynamic decomposition voltage of water is 1.23 V vs. RHE). Recently some solid and semisolid (gel) electrolytes (that acts both as the electrolyte and separator membrane) are also employed to configure the flexible/stretchable SCs. During operation (charging/discharging), a perfect synergy among all the components is strictly required to get better charge storage efficiency

© The Author(s), under exclusive licence to Springer Nature Singapore Pte Ltd. 2018   5
S. Ratha and A. K. Samantara, *Supercapacitor: Instrumentation, Measurement and Performance Evaluation Techniques*, SpringerBriefs in Materials, https://doi.org/10.1007/978-981-13-3086-5_2

and long-term running of the supercapacitor cell. Therefore, extreme care must have to take during configuration of the components to design a complete SC cell for practical applications. Basing upon the requirement to power electronic devices, the supercapacitors are designed in both the rigid, stretchable, and flexible configurations that are discussed in detail in the following sections.

2.1 Different Configurations of Supercapacitor

In order to power the advanced electronics, the supercapacitors are configured into different forms. These configurations are carried out based upon the types of electrode used, selection of electrolyte, and other inactive components like current collectors, separator, casings/packaging materials. The different types of configurations are summarized in the following (Fig. 2.1).

2.1.1 Conventional Configurations

2.1.1.1 Three-Electrode Configuration

Generally the supercapacitor testing electrochemical cells are designed in two basic types as three-electrode and two-electrode configurations. The first one comprises only one working electrode (containing the active electrode material), a reference (Ag/AgCl or saturated calomel electrode, SCE), and an auxiliary (platinum wire) electrode, and the measurement is performed by dipping in a suitable electrolyte solution (may be aqueous or non-aqueous). This type of arrangement is inevitably used to investigate the chemistry of the active materials, i.e., process of electrochemical reactions on the electrode surface (whether it is a diffusion- or kinetic-controlled process) and to study the redox behavior of the active electrode material. But a huge

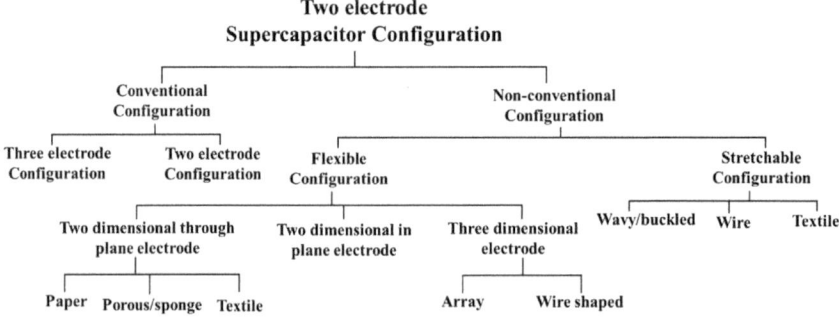

Fig. 2.1 Different type of possible configurations in supercapacitor

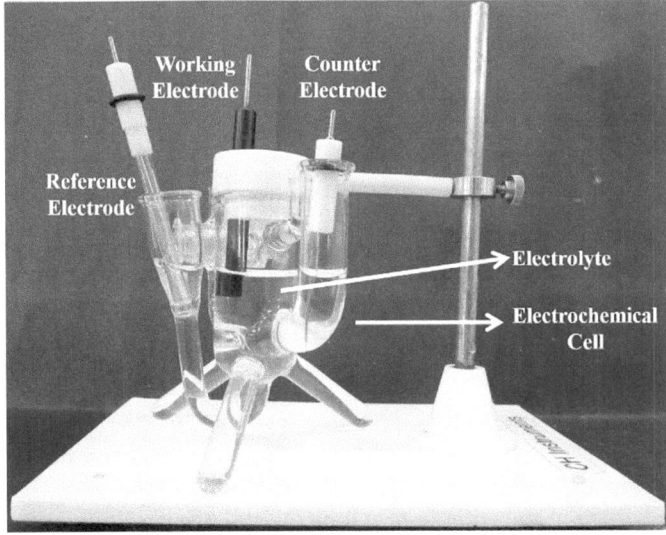

Fig. 2.2 Schematic presentation of a three-electrode arrangement used for the electrochemical measurement of an electroactive material for energy storage application

error has been realized when aimed to calculate the capacitance, energy storage, and power delivery efficiency of supercapacitor device using this type of electrode configuration. The following figure represents a three-electrode configuration generally used for the performance evaluation of active electrode material (Fig. 2.2).

2.1.1.2 Two-Electrode Configuration

On the other hand, in two-electrode configuration, a porous membrane (separator) soaked with electrolyte sandwiched between two working electrodes. The active materials are coated on the current metallic plates (current collector; foils of nickel, gold, stainless steel, etc.) to make the working electrode. Generally, in this arrangement, reference electrodes are not used, but in some cases it is inserted into the cell to monitor the actual potential change during the charging/discharging process. This model is a mimic of supercapacitor prototypes, and the measurement gives an actual value of (Fig. 2.3) or capacitance, energy/power values. These two-electrode configuration test cells are available in market in form of Swagelok type, metallic plate type with screws and coin cells, etc. One can make it in laboratory as per the required dimension.

These two types of electrodes are broadly used by the materials researchers to measure the efficiencies of the as developed material as well as the supercapacitor prototypes. But the applied potential and charge transfer reactions on the electrode surfaces of these two types of electrode configuration are completely different and

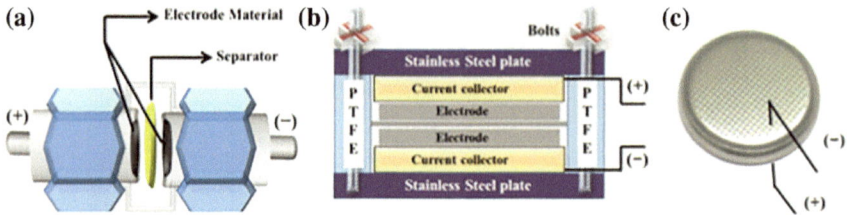

Fig. 2.3 Various types of two-electrode configurations used for the testing of supercapacitor performances

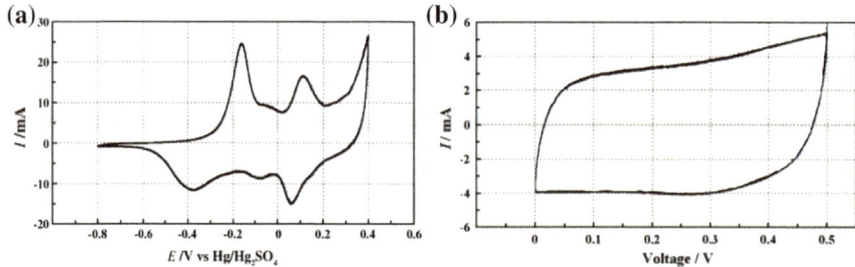

Fig. 2.4 CV of **a** three-electrode cell PANI/MWNTs and **b** a symmetric capacitor based on PANI/MWNTs composite electrode in 1 H_2SO_4 at scan rate, 2 mV/s (reproduced with permission from Khomenko et al. 2005)

can be seen from the following cyclic voltammogram pattern (Fig. 2.4) (Khomenko et al. 2005).

Based on the electrode materials used and mechanism involved for charge storage, the supercapacitors can be categorized as (i) *electrochemical double-layer capacitors* (EDLCs), (ii) *pseudocapacitors*, and (iii) *hybrid capacitors*. Generally the carbon-based materials (i.e., activated carbon, graphene, reduced graphene oxide, carbon nanotubes, carbon onions, nanohorns, etc.) are used in the EDLCs that stores charge electrostatically (by reversible physical accumulation of the charged ions at the electrode/electrolyte interface) (Futaba et al. 2006; Pandolfo and Hollenkamp 2006; Portet et al. 2008; Samantara et al. 2015; Samantara and Ratha 2018; Yang et al. 2007). Here the charge separation takes place by polarization at the electrode/electrolyte interface forming the double layer. Also a diffusion layer gets developed on the electrolyte site, which may be due to the accumulation of charged ions on the electrode surface. As a whole, the charge storage in the EDLCs is completely a surface phenomenon instead of that observed in batteries, thus providing higher power delivery values. Therefore, porous conductive materials possessing higher specific surface area are considered to be suitable one for EDLCs. Some reports mentioned that the pores of less than 0.5 nm are not accessible to the electrolyte ions and also the pores of size less than 1 nm are very small to accommodate the organic electrolytes (Kim et al. 2004; Qu 2002; Shi and Shi 1995). But the pore size and their distribution should be carefully optimized according to the electrolyte taken for the measurement. In this

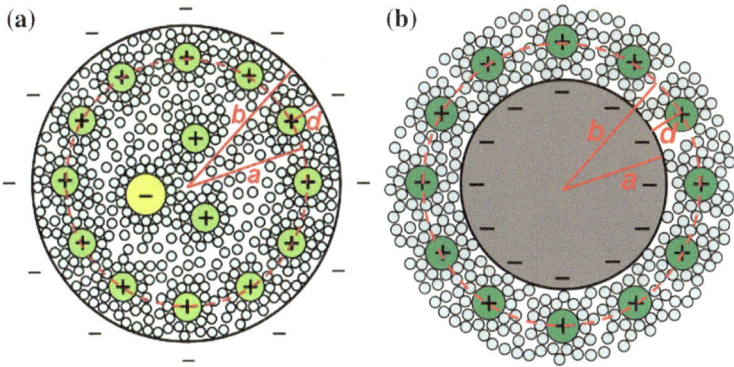

Fig. 2.5 a An endohedral capacitor with counterions close to the inner wall of a negatively charged pore inside nanoporous carbons such as activated carbons, template carbons, or CDCs. **b** An exohedral capacitor with counterions only residing on the outer surface of a negatively charged particle of carbon onions, end-capped CNTs, or carbon nanofibers (reproduced with permission from Huang et al. 2010, 2008)

regard, a carbon material having pores of 2−5 nm size (more than the size of two solvated electrolyte ions) are optimized to be suitable for the enhancement of energy and power of EDLCs.

Depending on the way of interaction of the counter ions to the carbon surfaces, these EDLCs are categorized as (a) *endohedral* and (b) *exohedral* capacitors (Huang et al. 2010, 2008). In the former type, the counter ions entered into the pores and form the double layer (Fig. 2.5a).

This type of charge accumulation has been observed in case of nanoporous carbons having negative surface curvature like activated carbons, carbide-derived carbon, and template carbons. Whereas in case of the exohedral capacitors, the ions observed to be reside on the outer surface of the carbon particles (Fig. 2.5b). The carbon materials having positive curved surfaces like end-capped carbon nanotubes, carbon nanofibers, and carbon onions show this type of behavior. Interestingly, owing to the zero curvature of graphene, it does not belong to either of these two categories of capacitor. On the other hand, the reversible redox reactions (electrolytic reactions) make the storage of charge in the pseudocapacitors. In this case, the redox active materials like metal oxides, sulfides, selenides, nitrides, mixed metal oxides, conducting polymers, and their composites with the carbon materials constitute the active electrode (Choi et al. 2006; MarriThese Authors Contributed Equally. et al. 2017; Ratha et al. 2017, 2016; Rudge et al. 1994; Samantara et al. 2018). These pseudocapacitors show a higher value of specific capacitance compared to that of carbon-based EDLCs. But due to involvement of redox reactions, like batteries, they show poor cycle performance.

Further the capacitors having similar electrode materials (either carbon or redox active materials) referred to as the symmetrical supercapacitors (SSCs). As these symmetrical SCs worked well, they are associated with lower capacitance and poor energy values compared to that of the traditional batteries. This may be due to their narrow operational potential window. For example, a carbon-based symmetrical SC used in aqueous electrolytes shows a maximum potential window up to 1 V owing to the decomposition of water and carbon oxidation. Therefore, each of the electrodes will work in a limited potential window leading to lower energy densities. The recent studies reveal that the energy densities of the supercapacitor can be increased either by (a) developing new electrode materials, (b) using non-aqueous or ionic liquid electrolytes, or (c) by integrating the carbon electrode with the pseudocapacitive/redox active electrodes. Hence, effort has been paid and the asymmetric arrangement of the electrodes has been introduced at the end of the 1990s. It based on the principle to use both the non-faradic EDL (as source of power) and faradic pseudocapacitive (as source of energy) electrodes with complementary potential windows in the same SCs cell. Here the charge storage takes place by following both the electrostatic and electrolytic reaction mechanisms. Nowadays, these hybrid supercapacitors come with two main approaches as (a) combination of pseudocapacitive electrode materials with the carbon-based materials and (b) lithium-inserted electrode with the carbon-based materials. Sometimes the second type of approach is regarded as the lithium-ion capacitor and mainly involves the use of lithiated metal oxides (PbO_2, NiOH, etc.) and carbon electrode to configure the cell (Burke 2007; Naoi and Simon 2008). Because of this type of arrangement, the capacitor cell performs in a wider working potential window showing higher capacitance values. By using this approach, 2–3 times enhancement in the energy densities has been realized compared to that of the conventional EDLCs.

There are various instrumentation or test methodologies that have been developed to evaluate the electrochemical charge storage performances of the electrode/electrolyte materials as well as designed supercapacitor prototypes. Out of which the cyclic voltammetry (CV), galvanostatic charge/discharge (GCD), and electrochemical impedance spectroscopy (EIS) are some of the general methods used (discussed in later sections). All these instruments are used to determine three major basic parameters, i.e., voltage, current, and time, from which the capacitance, energy/power densities, and equivalent series resistances can be derived.

2.1.2 Non-conventional Configurations

2.1.2.1 Flexible Supercapacitors

Owing to the requirements to power the wearable and flexible electronics, lightweight eco-friendly flexible/stretchable supercapacitors have been developed in a cost-effective manner. Further depending upon the arrangement of electrode materials and electrolyte/separator membrane, the flexible supercapacitors can be categorized as (a) two-dimensional through-plane electrode configuration, (b) two-dimensional in-plane electrode configuration, and (c) three-dimensional electrode configuration. All these are discussed in the following section.

(a) **Two-dimensional through-plane electrode configuration**: These are the flexible SCs having different types of 2D electrodes and can be subdivided into the following important types.

(i) *Paper-like Configuration*

In this type of the electrode configuration, the lightweight, flexible, and highly conductive materials like conductive carbon (carbon nanofibers, carbon nanotubes, graphene, graphene oxide, etc.), polymers (polyaniline), and their composites are used for the preparation of working electrode. Owing to the large surface area and short diffusion path (for electrons and ions), the carbon nanofiber (CNF) have been used as the substrate for grow of metal oxides ($NiCo_2O_4$ nanowires) (Huang et al. 2013). Further nanosheets of cobalt and nickel double hydroxides are electrodeposited on these NWs, and a higher value of areal capacitance (2.3 F cm^{-2} at 2 mA cm^{-2}), better rate stability (<40% decay by increasing current density from 2 to 150 mA cm^{-2}) with higher values of energy (~58.4 W h kg^{-1}), and power (~41.3 kW kg^{-1}) densities has been realized. Likewise, Xiao et al. developed a flexible conducting scaffold of $NiCo_2S_4$ nanotubes on carbon fiber and electrodeposited cobalt and nickel double hydroxides at different proportions (Xiao et al. 2014). Remarkably a higher value of areal capacitance (2.86 F cm^{-2} at 4 mA cm^{-2}) with a good rate capability (2.41 F cm^{-2} at 20 mA cm^{-2}) and better stability (4% loss even after 2000 repeated charge–discharge cycles at 10 mA cm^{-2} current density) has been observed. Further Cui's group have designed a flexible all solid-state supercapacitor using a carbon nanofiber network membrane (that acts as a free-standing electrode) and got higher volumetric capacitance (201 F cm^{-3} at 33 mA cm^{-3}) with 96% retention in initial capacitance even after 10000 repeated cycles (Sun et al. 2015). And 100% capacitance retention ratio has been observed for the designed SC device under continuous dynamic operations (bending and twisting). Not only the carbon fiber but also the carbon nanotubes are used for the designing of free-standing flexible SC electrode. These are generally prepared either by coating on the flexible sheets (office paper, acrylate sheets, etc.) or by forming CNT network films (by vacuum filtration method) (Kim et al. 2012; Xiao et al. 2013). But the high cost and complicated processes for preparation of stable CNT dispersion restrict its commercial application. Afterward the higher mechanical strength and better electrical conductivity of graphene motivate the energy researchers to use in energy applications. By using the flow-directed assembly, mechanical pressing of graphene aerogel, etc., the flexible graphene papers were developed and used in the preparation of flexible supercapacitors (Eda et al. 2008; Shao et al. 2015). These SCs devices observed to be showing better electrochemical charge storage and higher energy as well as power delivery efficacy compared to other carbon-based electrodes (Liu et al. 2014; Parvez et al. 2014; Zhang et al. 2012). Addition to these carbon-based materials, the polymers, metal oxides/sulfides/selenides, vanadyl phosphate, black phosphorous, and MXenes demonstrate as promising material for the preparation of flexible SC electrodes in different shapes as per the desirable applications (Acerce et al. 2015; Hao et al. 2016; Ling et al. 2014; C. Wu et al. 2013).

(ii) *Porous-/sponge-like configuration*

Highly porous–sponge-like materials have a great importance in the development of lightweight, flexible electrodes for SCs. It has been observed that the interconnected pores present in this sponge avail a short path length for the diffusion of electrons and ions facilitating the charge storage efficacy of the SC device. In view of these advancements, researchers have successfully substituted the polymeric scaffolds by developing graphene or carbon nanotube sponge having structure and pore distribution like the natural sponges. In a particular work, Chen's group have prepared the graphene foam having a three-dimensional network-like structure by simply carbonizing the melamine foam and observed a better capacitance value (250 F gm^{-1} in 1 M H$_2$SO$_4$ electrolyte at 0.5 A gm^{-1}). Further the composite of graphene foam with CNT shows better flexibility and good mechanical property demonstrating a promising substrate for designing of flexible SCs (Chen et al. 2013). Owing to the existence of large internal specific surface area, various metal oxides (MnO$_2$, Fe$_2$O$_3$, etc.), mixed metal sulfides (NiCo$_2$S$_4$), and polymers were developed on these carbonaceous sponge matrixes and successfully demonstrated as the free-standing electrodes for flexible SC devices. The scalable preparation of these binder-free composite can be carried out by following either the electrodeposition or hydrothermal processes (Chen et al. 2011, 2013; Li et al. 2012, 2013; Wu et al. 2013).

(iii) *Textile-like configuration*

Textiles are porous material having both the flexibility and stretchability and are obtained by simply weaving the natural and synthetic fibers. These configurations of SC have a great importance as the energy storage unit for wearable electronics. In this type of configuration, fibrous, flexible, and mechanically strong textile electrodes are employed in place of the conventional metal current collectors. These textile scaffolds are made up of plastics, metal, CNT, carbon fibers, graphene bundle, graphene yarn, etc. Not only as current collector but also these are used as free-standing electrodes in designing of flexible SCs. In these cases, the 3D network-like structure provides a fast conduction path for electrons/ions and has a higher mass loading capacity availing better electrochemical charge storage performances. For example, Cui et al. have prepared a conductive textile electrode by coating CNT on polyester fibers and then form a thin layer of MnO$_2$ on this flexible scaffold by electrodeposition process (Hu et al. 2011). In comparison to a flat metal substrate, these textile electrodes permit a higher mass loading of up to 8.3 mg cm^{-2} resulting in a higher areal capacitance value (2.8 F cm^{-2} at a scan rate of 0.05 mV s^{-1}). Afterward these fibrous polymeric supports of these textile electrodes were replaced by highly conductive carbon fibers, and a better performance has been achieved.

Using the activated carbon fiber cloth as the substrate, Dong et al. prepared CNT and MnO$_2$/CNT composites (Dong et al. 2016). A synergistic effect of carbon fiber cloth (it has higher electrochemical activity), CNT (better electrical conductivity), and MnO$_2$ (extremely high theoretical capacitance) has been observed in these composite materials that provides higher areal capacitance (2542 mF cm^{-2}), energy (56.9 μW h kg^{-1}), and power (16287 μW cm^{-2}) densities. Also much more work on these textile cloth-based electrodes and composites were carried out, and their charge storage performances were explored (Liao et al. 2015; Liu et al. 2017; Wang et al. 2015).

(b) **Two-dimensional in-plane electrode configuration**

These types of configured SCs are inevitably needed to power the miniaturized electronic gadgets. Moreover, the integration of these storage units should be very close to the electronic circuit to avail excellent nano/microscale peak power. In this regard, much more scientific efforts have been paid to develop the thin film technologies, optimization of size/shape of the electroactive materials as well as exploitation of noble device architectures to the SCs. Therefore, using the thin films of better electrical conductivity, higher electrochemical active surface area, and having very short electron/ion transport path, the micro-supercapacitors (m-SCs) were designed. By using the micro-fabrication process and in situ polymerization method, Wang et al. in 2011 have developed an all solid-state flexible m-SCs on the polyethylene terephthalate substrate (Wang et al. 2011b). These m-SCs comprise the array of polyaniline nanowires and show a volumetric capacitance of $588\,\mathrm{F\,cm^{-3}}$ with fast rate capability and lower current leakage value. Further by manipulating the structure and morphology of graphene, Yoo et al. have designed the in-plane SC device (Yoo et al. 2011). In this case, the reduced multilayer graphene oxide (synthesized by following the chemical vapor deposition method) employed as the electrode material and observed an aerial capacitance up to $390\,\mathrm{\mu F\,cm^{-2}}$. In another work, a graphene-based in-plane interdigital m-SC has been developed by Wu et al. on silicon substrate (Wu et al. 2013). The microelectrode patterns were developed by employing the lithography technique to the thin film of reduced graphene oxide (here the graphene oxide was reduced by using the CH_4 plasma) and show higher values of volumetric capacitance (up to $17.5\,\mathrm{F\,cm^{-3}}$) in H_2SO_4/PVA electrolyte with better energy ($2.5\,\mathrm{mW\,h\,cm^{-3}}$) and power ($495\,\mathrm{W\,cm^{-3}}$) densities. Nowadays instead of the high-cost conventional fabrication techniques, the m-SCs are prepared by the printing on the flexible substrates (PET, paper, etc.). Mullen's group have developed printable m-SC on both the paper and PET via a shadow mask using ink of the electroactive material and polymer (PEDOT: PSS) (Liu et al. 2016). The paper-based flexible m-SC shows an areal capacitance of $5.4\,\mathrm{mF\,cm^{-2}}$ with a very good rate capability (75% retention when scanned from 10 to $1000\,\mathrm{mV\,s^{-1}}$) demonstrating the best one among the graphene-based m-SCs (Fig. 2.6).

(c) **Three-dimensional electrode configuration**

It is well known that the charge storage in SCs is a surface phenomenon, so the three-dimensional electrode arrangement came into the picture, and more effort has been devoted to this aspect. This three-dimensional arrangement can be carried out in two main configurations like array and wire-shaped configurations as discussed in the following sections.

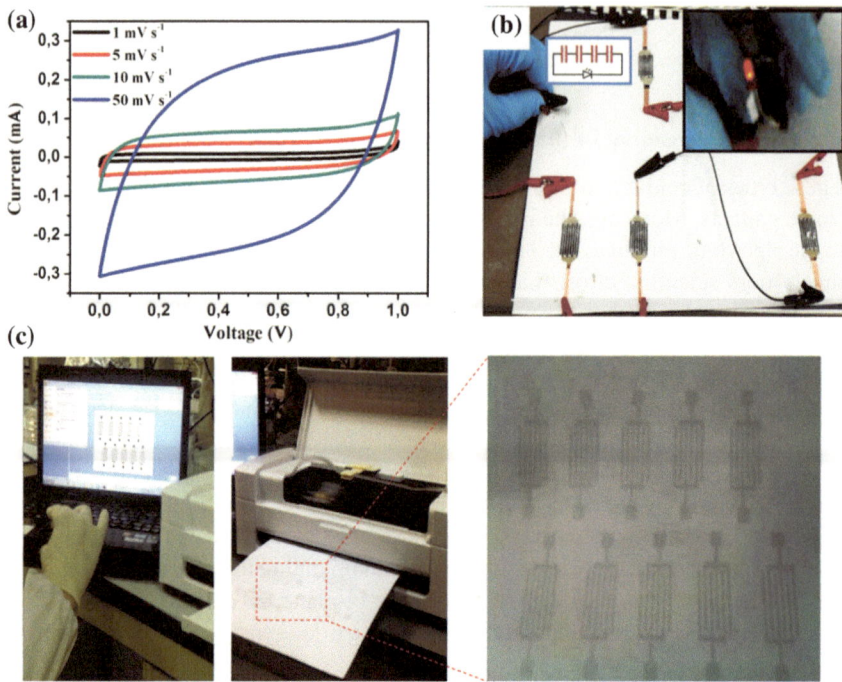

Fig. 2.6 **a** Cyclic voltammetry curves of an EG/PH1000 hybrid ink (total amount 10 mL) m-SC on a paper substrate at scan rates of 1–50 mV s^{-1}. **b** A charged m-SC array containing four single devices can power an LED. **c** Inkjet printing of custom-designed MSC arrays from a "home computer and printer" using pristine EG ink (reproduced with permission from Liu et al. 2016)

(i) *Array Configuration*

Flexible SCs having array of three-dimensional electrodes show an improved charge storage performance compared to the traditional SCs having planar electrodes. In these cases, the active electrode materials are prepared on the flexible electrocon-ductive substrates such as nickel foam, carbon cloth, polymer cloth, and yarn follow-ing different synthetic methodologies. In a specific work, Wang et al. have designed a flexible SC device using carbon cloth-supported single-walled carbon nanotube (SWCNT)/polyaniline composite material (as the working electrode) (Wang et al. 2011a). In the first step of electrode preparation, the SWCNTs are deposited onto the cloth by simply dip coating and then the nanowires of PANI are grown following a dilute polymerization process. These composite electrodes were assembled to form a flexible SC device and show a specific capacitance of 410 F g^{-1} with better cyclic stability. Not only the symmetric SC but also the asymmetric SCs are being devel-oped using these types of three-dimensional array-configured electrodes showing better charge storage performances even in the twisted and bent conditions (Xu et al.

2013). On the other hand, the excellent electrical conductivity, uniform microporosity and large accessible surface area; the nickel foam are assumed to be a suitable candidate for the development of flexible SCs. Using Ni foam as the substrate, Zhou et al. have developed CoO nanowire arrays uniformly coated with the polypyrrole (Zhou et al. 2013). The flexible SC prepared by assembling this electrode shows a specific capacitance of 2223 F g^{-1} with higher cyclic stability (99.8% retention of the initial capacitance value even after the 2000th cycle). Likewise the carbon cloth electrodes, the Ni foam-based flexible electrodes are also employed to design the flexible asymmetric SCs demonstrating very high-value specific capacitance, energy (43.5 Wh kg^{-1}), and power (5500 W kg^{-1}) densities.

(ii) *Wire-shaped Configuration*

These types of configured flexible SCs having one-dimensional wire-like electrodes show a potential candidature as a storage unit to power the wearable electronics and other miniaturized electronic gadgets. These SCs are also regarded as the fiber SCs and can be formed either by taking two parallel fibers, two twisted fibers, or one coaxial fiber. Wang's group from Georgia Institute of Technology has successfully developed the first fiber-based m-SC using ZnO nanowires as the active electrode on the Kevlar fiber substrate (Bae et al. 2011). Thereafter many works on the polymer-based fiber SCs have been presented. Besides these polymers and metal oxides, the flexible wire/fiber SCs have also been developed using carbon-based fiber scaffolds (made up of using graphene, CNTs, etc.) (Dong et al. n.d.; Xiao et al. 2012). Meng et al. have developed all graphene core–sheath fiber electrodes in which the core graphene fiber is covered by a sheath of three-dimensional highly porous graphene network (Meng et al. 2013). The all solid-state SCs was made by assembling these electrodes with H$_2$SO$_4$/PVA gel electrolyte. Because of the better conductivity of the core graphene fiber and large accessible surface areas of the porous graphene sheath, the SCs shows an excellent capacitance with better energy (0.4–1.7 × 10^{-7} W h cm^{-2}) and power (6–100 × 10^{-6} W cm^{-2}) densities. Peng's group have developed a wire SC using flexible fiber tube electrodes of hollow reduced graphene oxide/conducting polymer fibers (HCFs) and hollow reduced graphene oxide fibers (HPFs) (Qu et al. 2016). The all solid-state symmetrical SCs were prepared using two parallel HCFs electrodes in H$_3$PO$_4$/PVA electrolyte and observed a very high areal capacitance value of 304.5 Mf cm^{-2} (at 0.08 mA cm^{-2}) with high energy (27.1 μW h cm^{-2}) and power (66.5 μW cm^{-2}) densities. More interestingly, it shows constant charge storage performances even in different deformations. As is well known that during the practical applications, these flexible SCs will undergo various types of unavoidable mechanical deformations and the SCs should not have to lose its efficiency during this operation. In this regard, Huang et al. for the first time developed a shape memory supercapacitor (SMSC) using MnO$_2$ and polypyrrole on the substrate of Nickel–Titanium alloy (Huang et al. 2016). The SMSC shows better performances at different bending and twisting conditions demonstrating longer life span for practical applications.

2.1.2.2   Stretchable Supercapacitors

(i)   *Wavy/buckled Configuration*

The wavy, buckled, and wrinkled electrode configurations are widely adopted to structure the stretchable SCs (s-SCs). Yu et al. have reported a pioneer work on stretchable SCs using the buckled-structured electrodes (Yu et al. 2009). A multi-step process has been adopted for fabrication of electrode; in the first step, the buckled shape of SWCNT was made on the functionalized surface of pre-stretched poly (dimethyl) siloxane slab followed by the irradiation of UV-light in the second step. Then a periodically buckled surface has been observed by removing the strain on the pre-strained slab. No such significant change in the CV pattern and capacitance value has been observed even on application of 30% strain on this stretchable SC device. Though it worked well, the applied strain was limited, i.e., up to 30% restricting the practical application. Also the decay in capacitance values due to the leakage of liquid electrolyte and the lamination between the electrode and electrolyte may cause the decrease in the performance of the stretchable SCs. In order to eliminate these limitations, various stretchable SCs are made using the gel electrolytes (Niu et al. 2013; Xie et al. 2014). Other carbon materials like graphene are used to prepare the wrinkle configured s-SCs. In a particular work, Zhang et al. have prepared the crumpled graphene papers by transferring graphene papers to the pre-strained elastomeric substrates and observed a high specific capacitance of 196 F g^{-1} and better reliability over 1000 stretch/release cycles (Zang et al. 2014). Qi et al. have developed highly stretchable micro-SCs using tripod-structured graphene micro-ribbons (Qi et al. 2015).

The authors proposed two main advantages for this particular configuration, i.e., (i) the electrode material in the m-SCs are more stretchable instead of showing intrinsic stiffness as in planar configurations and (ii) provides a reduced strain on the electrode fingers during the stretching/relaxing processes of the m-SCs electrodes. At first, the graphene oxide (GO) micro-ribbons were obtained by immersing the lithographically patterned Cu films in the GO suspension (Fig. 2.7). Then a layer of Au film was deposited that acts as the current collector followed by coating with a very thin layer of PDMS. The prepared film was placed on a pre-strained tripod-shaped PDMS substrate and subsequently removed the Cu film and reduced the GO layer (treating with hydrazine hydrate) to reduced graphene oxide (rGO). Finally the micro-ribbons of graphene were assembled to form the m-SCs using PVA/H$_3$PO$_4$ gel electrolyte, and stable charge storage performances have been demonstrated under different strains (0–100%). This type of configuration finds a potential application as a storage device in stretchable and wearable electronics.

(ii)   *Wire-like configuration*

The wire-like configured electrodes are able to afford higher strained deformations to the s-SCs. For the first time, Peng's group has developed a stretchable fiber-shaped SCs showing enhanced electrochemical and mechanical properties. They have wrapped the CNT sheets onto an elastic fiber and finally assembled the s-SC

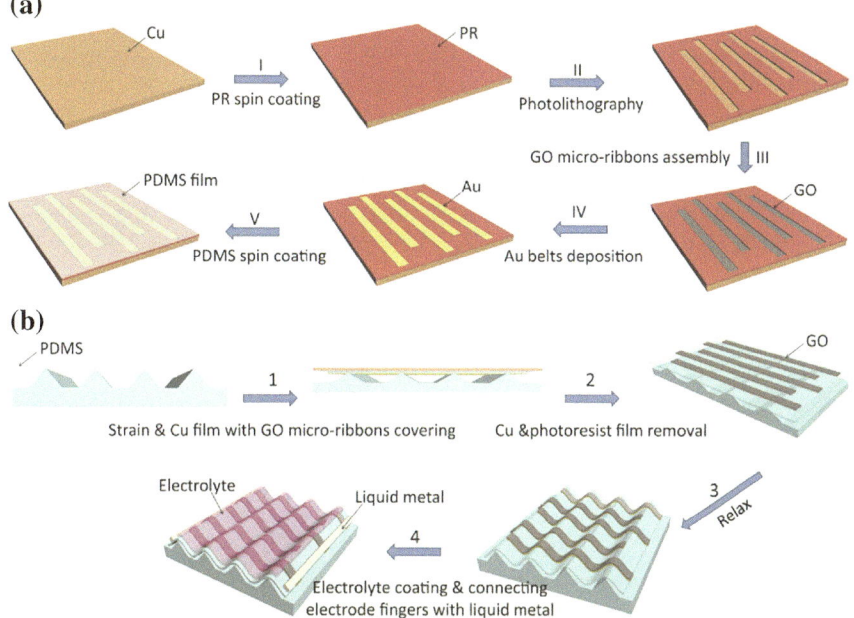

Fig. 2.7 Schematic drawing of fabricating process of the stretchable micro-supercapacitor. **a** Fabrication of the GO electrode arrays by photolithography, synchronous reduction, assembly strategy, and thermal evaporation processes: (I) A layer of photoresist was spin coated on the Cu foil; (II) patterning the Cu foil by photolithography in the form of desired interdigital patterns; (III) GO micro-ribbons assembly via a synchronous reduction and assembly strategy; (IV) a thin layer of gold film was covered on the sample surface by thermal evaporation via shadow mask; and (V) a thin layer of PDMS was spin coated on the sample surface. **b** Transferring electrode arrays onto the tripod-structured PDMS substrate: (1) stretching the tripod PDMS substrate and the sample with a layer of half-cured PDMS film was pasted on the substrate surface; (2) removing the Cu foil and photoresist with FeCl$_3$ solution and acetone, respectively; (3) releasing the pre-stretch to get the devices with suspended wavy-structured electrode arrays; (4) coating the electrode array with electrolyte (H$_3$PO$_4$/PVA) and connecting the electrode fingers with liquid metal (reproduced with permission from Qi et al. 2015)

using gel electrolytes. It shows capacitance up to 41.4 F g^{-1} with 95% retention after 100 repeated cycles under 75% strain. Later on, various fiber-shaped SCs were developed using CNT, CNT/metal oxides, and CNT/polymer composites demonstrating potential application under various percentages of stretchability (Choi et al. 2015; Zhang et al. 2015). By combining the buckled configuration and wire-like structure, Liu et al. have prepared a sheath–core-structured stretchable fiber SC (Liu et al. 2015). At first the CNT sheets (oriented in wire direction and acts as sheath) wrapped onto a pre-stretched rubber wire (1400% strain) and obtained a buckled wavy-structured CNT layer by releasing the strain. The resulted supercapacitor shows a reversible stretchability in both the axial as well as in the belt directions showing <5% resistance change under 1000% applied strain enabling suitability in various applications.

(iii) *Textile Configurations*

Like in case of the textile-configured flexible SCs, here also the conductive and stretchable fiber coated with active electrode material (CNTs and other carbon-based materials, metal oxides, polymers, etc.) are used to assemble the s-SCs. In some cases also a better stretchability has been obtained by weaving both the elastic and inelastic fibers in an alternative manner. Lee et al. have developed a tricot-weave textile from the polyester (elastic fiber) and Spandex (non-elastic fiber) yarns and coated with multi-wall carbon nanotubes (MWCNTs) (Lee et al. 2015). The s-SC thus resulted from this MWCNT-coated electrodes shows an uninterruptable capacitance ($35\,\mathrm{F\,g^{-1}}$ at $0.25\,\mathrm{A\,g^{-1}}$) value under application of different strains. Also this charge storage performance observed to be well retained even on application of 50% strain in biaxial direction (Fig. 2.8). Also the s-SCs have designed by screen printing the conducting electroactive inks onto a pre-designed stencil. They show better charge storage performances under very high level of stretchability demonstrating their potential for practical applications (Liao et al. 2015; Wang et al. 2015).

2.2 Key Parameters

2.2.1 Specific Capacitance

The capacitance (C) is one of the important key parameters, i.e., the amount of charge stored over a change in voltage and generally used to evaluate the performance of a supercapacitor. The energy researchers are always trying to explore new electrode materials for better charge storage performance of the SCs. For this, one has to calculate the intrinsic capacitance of each of the materials and to perform a comparative study. So it will be easy to decide whether to go for that particular material or not. This intrinsic capacitance of electrode material is normally presented in terms of specific capacitance (C_s) and can be obtained by normalizing the observed capacitance with mass (gravimetric capacitance; $\mathrm{F\,g^{-1}}$), area (areal capacitance; $\mathrm{F\,cm^{-2}}$), length (linear capacitance; $\mathrm{F\,cm^{-1}}$), or volume (volumetric capacitance; $\mathrm{F\,cm^{-3}}$) of electrode material. Further the performance of the complete SC device can be evaluated by dividing the calculated capacitance value by whole mass/volume of the device. It has been observed that the value of C_s can be affected by a number of factors like mass loading (mass of active materials per unit area of the electrode), dimension, thickness of active material, electrode density (the amount of active material packed per unit volume of the current collector), types of electrolyte/separator used, and the experimental conditions. Therefore, one has to optimize all these parameters to observe the intrinsic capacitance of a particular material. The value of C_s as well as C can be calculated from either recording the cyclic voltammograms, galvanostatic charge–discharge or from the equivalent series resistance that are elaborately discussed in the latter part of this book.

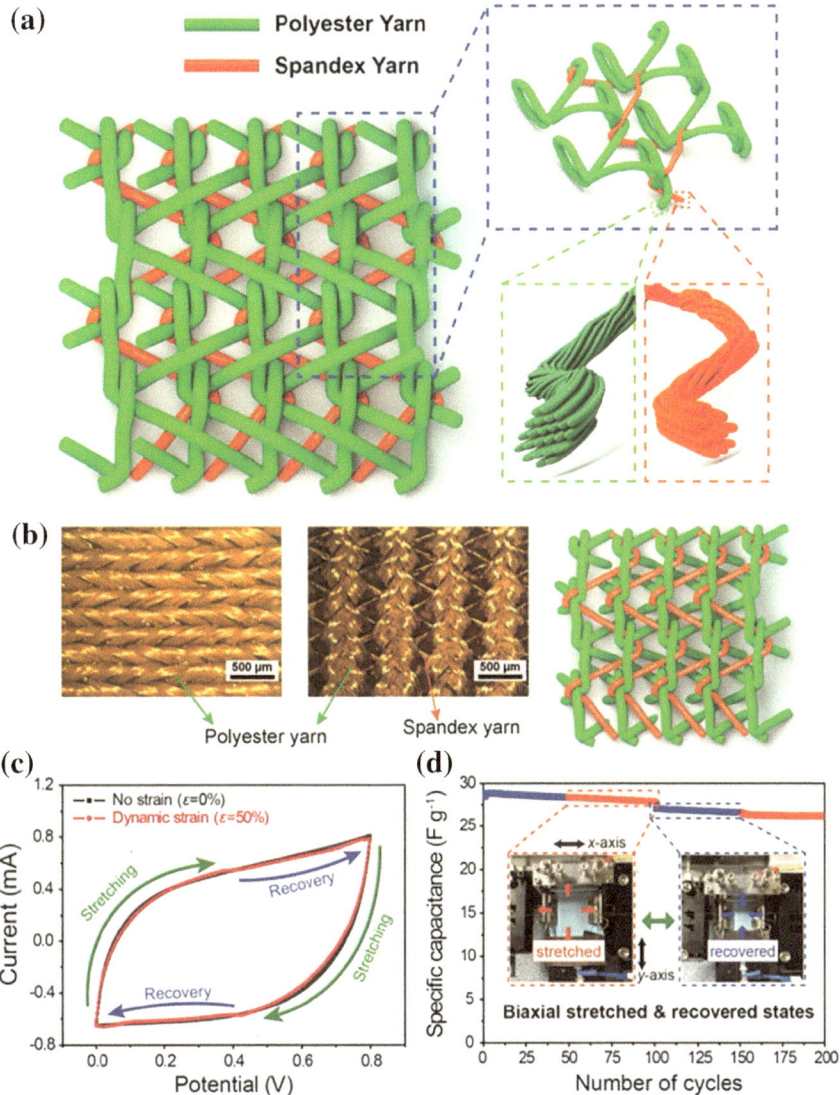

Fig. 2.8 a Schematic illustration of the tricot-weaving structure consisting of two kinds of yarn (green: polyester yarn, orange: spandex yarn), a focused view of the weaving structure made of both yarns, a further zoomed-in view of the weaving point to illustrate that each yarn consists of a bundle of fibers. **b** Optical microscopy images of the actual textile electrode from the front and the rear sides after Ag nanoparticles coating, along with a rear view schematic. **c** CV curves under and without dynamic strain at a scan rate of 80 mV s^{-1}. **d** The cycling performance under biaxially 50% strain and recovered states (reproduced with permission from Lee et al. 2015)

2.2.2 Specific Energy and Specific Power

These are the most important parameters for every energy storage/conversion systems and need to be evaluating to validate their practical applications. Energy of a storage/conversion system is the amount of electrical energy stored in or released from the SC device and can be calculated by integrating the charging or discharging curve, respectively. Further the ratio of the energy stored to energy deliver gives the idea regarding the energy efficiency of that particularly designed supercapacitor cell. The value of this energy can be derived from the capacitance and presented in terms of specific energy by normalizing either with respect to the mass loaded on the electrode (gravimetric energy; W h kg^{-1}) or with the volume of the electrode (volumetric energy; W h L^{-1}). Whereas the value of specific energy strongly depends on the capacitance and operating potential window. Therefore, much more effort is given to develop more efficient electrode material and to explore various electrode configurations (like asymmetric SCs, hybrid SCs, lithium-ion capacitors).

On the other hand, the rate at which the energy transfer takes place from or to the SC is termed as power and represented in terms of specific power by normalizing with either the mass loading (gravimetric power; W kg^{-1}) or with the volume (volumetric power; W L^{-1}) of the SC electrode. This specific power of a SC can be determined from the capacitance values. Additionally other methods are also been developed, i.e., the pulse energy efficiency method (PEE), IEC-62576 [IEC-62576, 2006], and DOE-Freedom Car [DOE/NE-ID-11173, 2004] method to calculate the actual specific power of SCs (Burke and Miller 2011; "Electric double layer capacitors for use in hybrid electric vehicles—Test methods for electrical characteristics," 2006; Idoho National Laboratory 2004). It has been reported that SCs have more than 100 times specific power compared to that of the fuel cells and traditional lithium-ion batteries. Therefore, the SCs are considered as the better substitute for batteries as the advanced energy storage units for various applications. The comparison of specific energy and power of different electrochemical energy storage systems are represented in Ragone plot (as shown in Fig. 2.9) (Zhang and Pan 2015).

2.2.3 Operating Potential Window

It refers to the potential applied to the SC or the suitable potential window within which the SC operates safely. The potential window (P_w) for a particular SC device or electrode material strongly depends on the types of solvents used for the preparation of the electrolyte (whether it is aqueous or organic) taken for that measurement. Both of the cyclic voltammograms and the galvanostatic charge–discharge techniques are used for the optimization of P_w. It has been reported that in case of the aqueous electrolytes, the value of P_w is limited to 1.0 V (since the thermodynamic decomposition of water takes place at 1.23 V vs. RHE) and varies between 2.3 and 2.7 V for the non-aqueous electrolytes (Kurzweil and Chwistek 2008; Merlet et al. 2013; Saman-

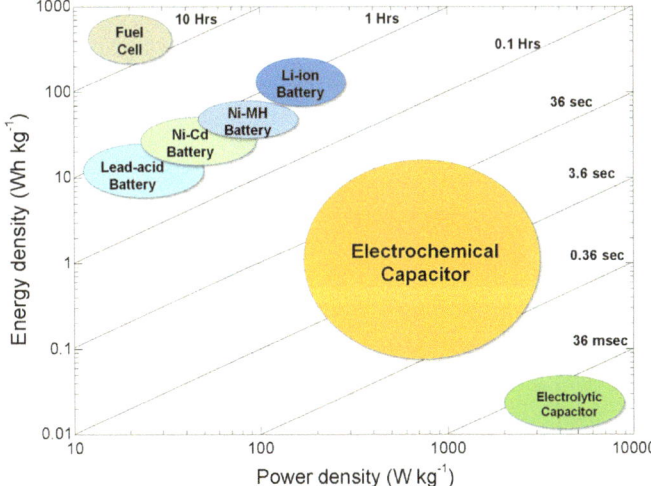

Fig. 2.9 Ragone plot presenting the power and energy densities for several electrochemical energy storage devices (reproduced with permission from Zhang et al. 2015)

tara et al. 2018). Since values of all the major parameters, i.e., the specific energy and specific power are strongly depending on the value of this P_w, so it is necessary to optimize the electrolyte so that it will sustain in a wider range of working potential. Further the electrodes can be designed in different configurations to get enhanced values of P_w. For example, in case of the asymmetric SCs, the pseudocapacitive materials are integrated with the carbon electrode and a higher P_w value has been realized even in aqueous solvent-contained electrolytes (Conway 1991; Gao et al. 2012; Qu et al. 2008). Also many researches are going on to increase the P_w values so that increased values of specific capacitance, specific energy, and power can be obtained.

2.2.4 Cyclic Stability

It is the major metrics of the SC device and is highly desirable to evaluate for their practical applications. It has been reported that the SCs are very robust in nature and runs continuously more than one lakh repeated cycles which is nearly ten times more than that of the traditional energy storage systems. Practically, calculation of such a higher value is a tedious task and therefore is measured by determining the capacitance retention rate. This retention rate can be derived by comparing the specific capacitance (either by using the cyclic voltammograms or galvanostatic charge–discharge technique) of the initial cycle to that of final cycle of the SC and is presented in a plot of time against the percentage of capacitance retention. In a particular work, Uno and Tanaka have performed the retention rate measurement of SCs for 3.8 years and observed a linear decrease in the capacitance retention rate with respect to the square root of the number of cycles (Uno and Tanaka 2012).

Chapter 3
Characterization and Performance Evaluation of Supercapacitor

Characterization of supercapacitor electrodes/devices primarily involves both cyclic voltammetry and constant current charge–discharge techniques. Nevertheless, considering the induction of large number of non-carbonaceous materials to fabricate supercapacitor electrodes, additional characterization tools are required that are going to be detailed in the succeeding sections. Since diverse characterization techniques are being implemented to address all the electrochemical processes involved in the supercapacitor electrode (or device) measurement (or evaluation), it is highly imperative to have at least a brief account of these processes taking place at the electrode/electrolyte interface.

3.1 Double-Layer Formation and Diffusion Process

In conventional capacitors, the two electrodes are separated by an insulating material (or dielectric), and the charge accumulation occurs through electrostatic attraction between two oppositely charged particles or temporary polarization induced in the dielectric itself. Though electric double-layer capacitors are assumed to follow a similar principle, the charge accumulation is achieved through oppositely charged ions forming inner and outer Helmholtz layers. The use of an electrolyte instead of dielectric not only provides a highly conducting channel but also creates an ion-rich environment facilitating both the movement of electrons and ions. In principle, the formation of inner Helmholtz layer is supposed to occur via electrostatic force. However, surface of EDLC materials such as activated carbon or reduced graphene oxide contains both structural and chemical irregularities and the adsorption process initiates with electrochemical interaction between the charged species (inner Helmholtz layer) with subsequent formation of outer Helmholtz layer via coulombic force (electrostatic) between oppositely charged ions (Eftekhari and Garcia 2017). Here, the non-uniformity of the electrode surface does not allow the uniform distribution of surface adsorbates, and if they are arranged randomly with favorable

© The Author(s), under exclusive licence to Springer Nature Singapore Pte Ltd. 2018   23
S. Ratha and A. K. Samantara, *Supercapacitor: Instrumentation, Measurement and Performance Evaluation Techniques*, SpringerBriefs in Materials,
https://doi.org/10.1007/978-981-13-3086-5_3

sites occupied first, then surface diffusion process is bound to play a crucial role in preventing the degradation in the charge accumulation process in supercapacitors. Not only carbon materials with high surface areas, but other electrodes derived from carbon-based starting materials also show a myriad of supercapacitor performances due to different synthesis techniques adopted and/or the presence of surface defects, functional groups, dangling bonds, and dopants, etc. (Eftekhari 2018).

The process of surface diffusion, however, requires negligible potential energy to occur and has been found to play a significant role in the overall charge storage performance of a supercapacitor electrode/device (Eftekhari 2018). This process can actually trigger additional surface reactions (faradic reactions) in EDLCs which are difficult to distinguish considering the fact that both the processes of double-layer formation and surface reaction occur at almost the same timescale. These faradic reactions are more or less equivalent to low energy chemisorption and are fast enough to proceed along with the double-layer charging. However, at higher sweeping potential values (e.g., at 200 mV/s onwards), the faradic contribution is largely suppressed as the change in current/potential is too fast for the surface reaction to catch up.

3.2   Surface Adsorption

In the previous section, we have discussed the faradic contribution of EDLC materials (though negligible) that distinguishes them from ideal double-layer materials let alone the potential current response. However, there are materials composed of a core metal atom with wide oxidation state capable of producing very high background current mimicking the electrical response of EDLCs. The charge storage mechanism in these materials is predominantly faradic with small contribution from the physical adsorption process. The chemical adsorption (chemisorption) is one of many faradic processes and is much faster (sometimes can be highly reversible too) than the rest. However, the process of chemisorption involves the possibility of formation of chemical bonds (charge transfer between the ions and electrode material which can be investigated through XPS technique), thus requiring an energy threshold to proceed (Tõnisoo et al. 2013). This makes the whole process of chemisorption dependent on the applied potential unlike EDLC, where the adsorption process is independent of the applied potential (Eftekhari 2018).

The electrochemical adsorption process differs from the redox reactions in terms of activation energy. Also, the peaks obtained during the process of electrochemical adsorption are generally broader in comparison to those observed in redox reactions and give an impression of a capacitive behavior. So, carbonaceous materials may be termed as capacitive-like rather than capacitive.

3.3 Pseudocapacitors

If the whole process of charging and discharging can be made slower (as in the case of batteries), then the energy density of a supercapacitor device can be enhanced. However, there are subtle technical ways to carry out such modifications. Some materials are intrinsically redox active and some just rely on specific material designing. Either way, the underlying goal is to promote bulk diffusion rather than only surface diffusion (Eftekhari and Mohamedi 2017). Clearly enough, the bulk diffusion process would take more energy (and more time) than surface adsorption, and the process may or may not be reversible. The bulk diffusion (or more likely the solid-state diffusion) is more or less a battery concept rather than a supercapacitor. If a single-crystalline material is considered, then it may show a prominent redox peak (at a standard potential value) as it would have a preferential/uniform lattice growth limiting the redox active sites to that plane only.

However, if the same material is subjected to defects (of any kind), or else a polycrystalline form of the same is taken, then the standard redox peak may be accompanied by an underpotential redox reaction peak (surface adsorption), and another overpotential redox peak. Here, the characteristic redox peak of the material at standard potential shows a current value almost similar to the accompanied peaks, and if combined, they would mimic the current response of an EDLC material. However, pseudocapacitive materials do not possess high surface area as observed in the case of most of the carbon-derived electrode materials. Hence, it is highly unlikely for these pseudocapacitive materials to store large amount of charge via adsorption process alone. A schematic representation of the surface redox process in a pseudocapacitive material constituting underpotential, standard, and overpotential redox peaks has been illustrated in Fig. 3.1.

When electrode voltage is proportional to surface coverage and surface coverage is proportional to state of charge; i.e., current is linearly proportional to the applied potential, and the process is pseudocapacitive. Materials such as MnO_2 and RuO_2 have been widely accepted as pseudocapacitive properties as they rely on their excellent redox properties to generate i-V response similar to EDLC materials. However, in reality, they just produce a series of redox peaks at different potential values which are too close to form a rectangular plot making it difficult to distinguish the charging process. There have been reports on several metal oxides (especially oxides of transition metals) claiming to show enhanced capacitive behavior. This, however, is not true and can be highly misleading, as the underlying principle of a capacitor is to separate charge without assorting to adsorption or redox reactions. Also, the cyclic voltammetric response of a capacitor would give a constant value of capacitance over the whole potential range which is not true for the cyclic voltammograms obtained from pseudocapacitive materials (Forghani and Donne 2018a). The terminology, pseudocapacitance, therefore should be limited to materials having, at most, rapid surface redox properties (with highly reversible charging–discharging characteristics). Being said that a pseudocapacitive material could, under few circumstances, show battery-like behavior and vice versa.

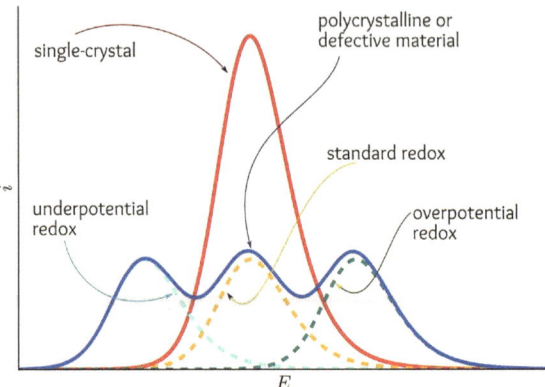

Fig. 3.1 An illustrative model of the voltammetric behavior of a metal oxide where some under-potential and overpotential redox sites exist along with the standard redox sites. The red curve represents a perfect single-crystalline material in which all the redox sites have the formal redox potential. The solid blue line depicts a case where only 1/3 of the redox sites react at the standard potential, and 1/3 reacts at $\times V$ below the standard potential (in the underpotential region) and 1/3 at $\times V$ above the standard potential (in the overpotential region). The dashed lines indicate the individual voltammograms where only the corresponding redox sites were active in the system (reproduced with permission from Eftekhari and Mohamedi 2017)

There is a contrasting difference between a pure faradic electrode and a pseudocapacitive electrode though they both might store charge via redox reactions (Conway et al. 1997). Pseudocapacitors are fast and even if there is any occurrence of redox reaction, the timescale is much smaller than that of pure faradic (battery) materials. As discussed above, the term pseudocapacitance may appropriately be associated with materials such as MnO_2 and/or RuO_2 having capacitive-like current response. Materials like $Ni(OH)_2$ and cobalt oxides should rather be termed as battery materials as their current response is completely nonlinear with the applied potential sweep (Brousse et al. 2015). There are rare instances where a pseudocapacitive material would exhibit battery-like intercalation process during charging besides fast surface redox reaction (Eftekhari and Mohamedi 2017). Such materials can be addressed by separately characterizing the two processes through specific measurement parameters. Also, there are reports where battery materials can produce capacitive-like i-V response, but it can only be correlated to the material designing rather than being specified as an intrinsic property of the material.

Figure 3.2 shows the actual mechanism behind the rectangular CV plot obtained from a true pseudocapacitive material due to the continuum between under potential, standard, and overpotential redox peaks.

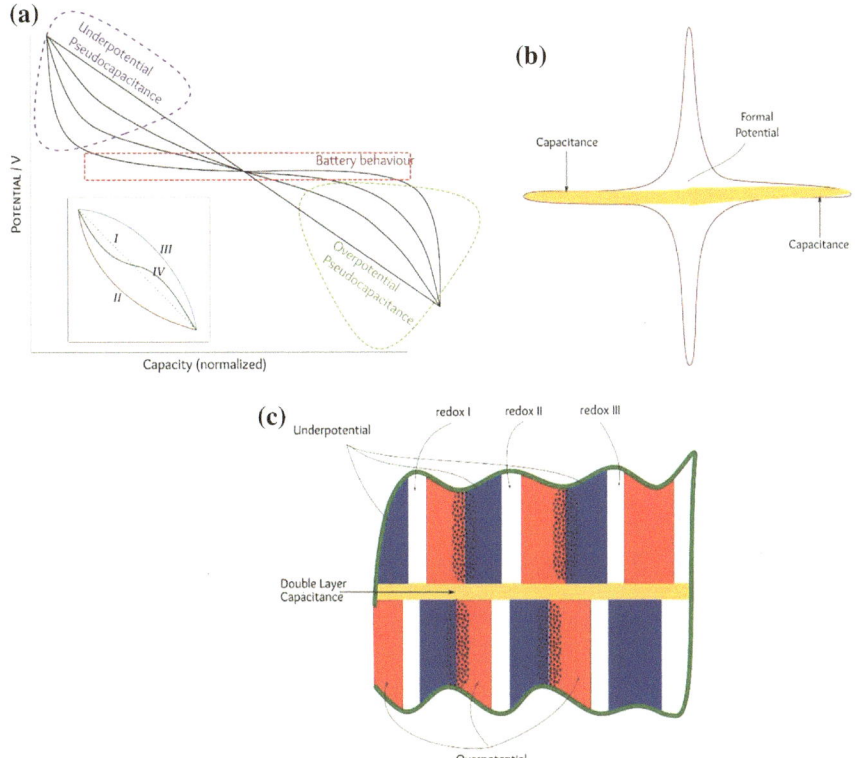

Fig. 3.2 Transition from battery to pseudocapacitive behavior. **a** Charge–discharge profiles: The *x*-axis has been normalized for the maximum capacity achieved in each case, **b** cyclic voltammogram of a battery material, **c** illustrating the contributions of under potential and overpotential processes in respect to the redox systems, creating a capacitive-like behavior. In this case, three redox systems are shown in which the peak is broadened by underpotential and overpotential regions (reproduced with permission from Eftekhari and Mohamedi 2017)

3.4 Evaluation Techniques

Now that we have an apparent understanding of the categorization of reported super-capacitor materials depending on their charge storage mechanisms, and the next thing that should be carefully done is their proper characterization. This can be done mostly through the observation of three basic parameters, i.e., the net capacitance of a supercapacitor device, working potential window, and the effective series resistance (ESR). It is to be noted that here the term net capacitance represents practical devices (two-electrode systems comprising symmetric, asymmetric, and/or hybrid electrode materials) rather than electrochemical half cells (three-electrode config-urations). Though essential in revealing the true characteristics (whether faradic or double-layer) of a single electrode and also the safe working potential window, device evaluation through three-electrode configurations could actually impart significant error due to high sensitivity.

The measurement techniques in both two- and three-electrode configurations are almost similar. Nevertheless, comparison of the subsequent mathematical calculations in both the cases could lead to ambiguity, for there are misinterpretations where authors have tried to correlate data obtained from both configurations. This could lead to severe misunderstanding of the appropriate use of the two types of configurations. A typical three-electrode configuration comprises a working electrode (to be measured/characterized), a suitable reference electrode (depending upon the electrolytic environment), and a counter electrode. The role of the counter electrode, in general, is only to complete the electrical circuit and its role in electrochemical activities is rather diminutive. It is assumed that the three-electrode system represents an isolated single electrode of a two-electrode device whose another electrode is at an infinite separation. Therefore, the concept of energy density and power density are both obsolete though many researchers have reported otherwise.

There have been attempts to replace the counter electrode with another electrode either similar (symmetric configuration) to or different (asymmetric or hybrid configuration) than the working electrode and characterize the whole system as a practical two-electrode supercapacitor device. In this case too, the high solution resistance, zeta potential, and sluggish cationic movements due to solvation tend to impart a semi-infinite separation between the two electrodes making the whole evaluation process erroneous.

3.4.1 Cyclic Voltammetry

The general concept of cyclic voltammetry is the application of a linearly varying potential (with an optimized range, called the working potential window) across the two electrodes of a two-electrode device, or between the working and reference electrode in a three-electrode system. The variation in the applied potential is achieved through an incremental potential step per second called as the sweep rate. Cyclic voltammetry in a three-electrode configuration is considered best practice to gather information regarding the charge storage occurring in supercapacitor devices to distinguish between a double-layer and pseudocapacitive material. Figure 3.3 illustrates a detailed schematic of the difference between capacitive, pseudocapacitive, and faradic in terms of both cyclic voltammetry and constant current charge–discharge techniques (Sarkar et al. 2018).

However, there are pseudocapacitive materials which have almost similar cyclic voltammograms as that of a double-layer material. Thus, cyclic voltammetry technique alone is not sufficient to quantize the percentage contribution to both double-layer and pseudocapacitive processes. As discussed earlier, a more accurate approach could be to check whether the current response is linear with the varying potential step or not. A linear relationship would suggest dominant double-layer mechanism while finite diffusion process at the electrode/electrolyte interface can be stated from

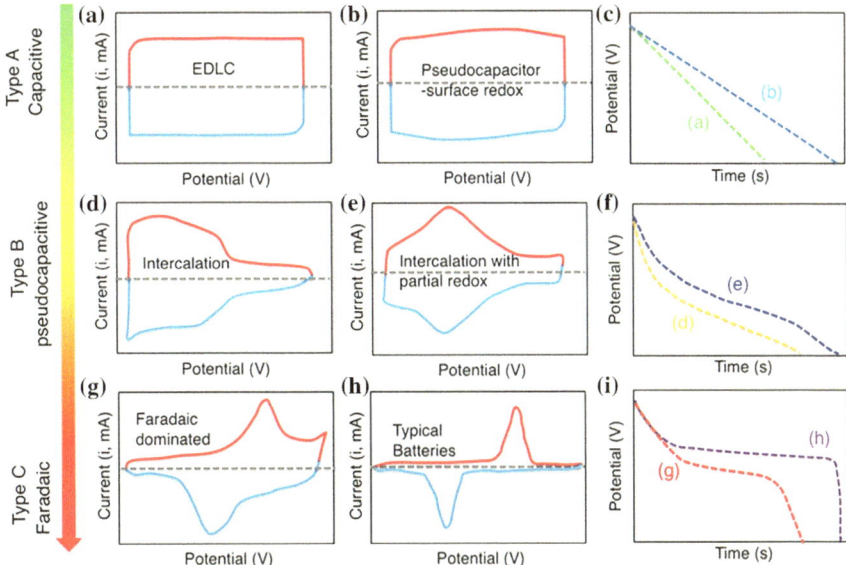

Fig. 3.3 (**a, b, d, e, g, h**) Schematic cyclic voltammograms and (**c, f, i**) corresponding galvanostatic discharge curves for various kinds of energy storage materials. A pseudocapacitive material will generally have the electrochemical characteristics of one, or a combination of the following categories: **b** surface redox materials (e.g., MnO_2 in neutral, aqueous media), **d** intercalation-type materials (e.g., lithium insertion in Nb_2O_5 in organic electrolytes), or **e** intercalation-type materials showing broad but electrochemically reversible redox peaks (e.g., Ti_3C_2 in acidic, aqueous electrolytes). Electrochemical responses in (**g–i**) correspond to battery-like materials (reproduced with permission from Sarkar et al. 2018)

slightest of deviation from the linear dependence of current on the potential sweep rate. However, both the double-layer and faradic processes occur at almost the same timescale making them practically indistinguishable. *CV* technique is also useful in optimizing the working potential window which depends on the electrolyte material, active electrode material, and several other key parameters (Samantara and Ratha 2018).

3.4.2 Step Potential Electrochemical Spectroscopy

Besides cyclic voltammetry, step potential electrochemical spectroscopy (SPECS) is another technique which can provide valuable insights regarding both double-layer and diffusion controlled charge storage processes in an electrochemical cell. The advantage with SPECS is that the contrasting charge storage processes can actually be calculated separately which would help understands the nature of the

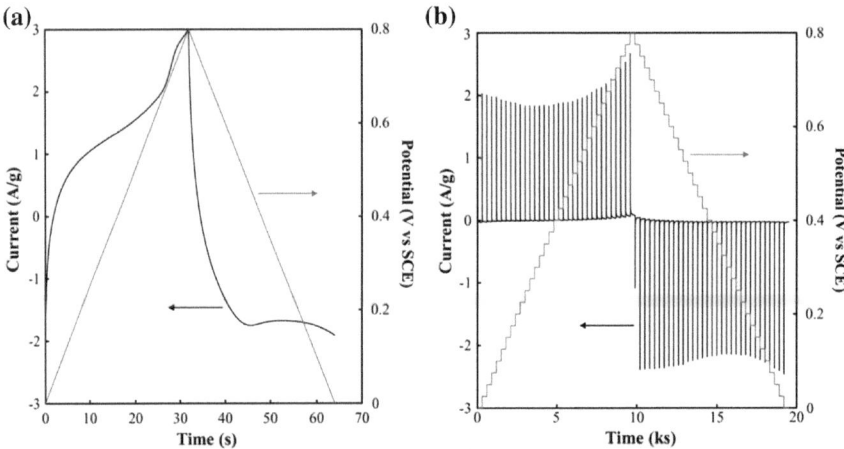

Fig. 3.4 Current response for **a** a *CV* experiment (EMD in 0.5 M K$_2$SO$_4$) cycled in the range 0.0–0.8 V versus SCE at a scan rate of 25 mV/s, and **b** for a SPECS experiment (EMD in 0.5 M K$_2$SO$_4$) with ±25 mV potential steps and a 300 s equilibration time (reproduced with permission from Forghani and Donne 2018b)

electrode material. SPECS is basically a time-dependent current response on applying a potential step.

In this method, a series of potential steps of equal magnitude are applied to the working electrode allowing for sufficient dwelling time during which a state of quasi-equilibrium is expected to form. However, diffusion processes, being on the slower side as compared to double-layer charging, do not agree well with the equilibrium state and leave the electrode surface with residual charge (or current). Therefore, SPECS method could be a handy tool in determining double-layer and diffusion-assisted charge storage along with residual current. More details on this technique and relevant parameters can be found elsewhere (Dupont and Donne 2015; Forghani and Donne 2018b; Gibson and Donne 2017). SPECS can reveal the contribution from all the underlying charging processes (i.e., double-layer, diffusion, faradic) as can be observed from the illustration provided in Fig. 3.4.

3.4.3 Constant Current Charge Discharge (CCCD)

In this technique, continuous charging and discharging is carried out for the super-capacitor device or a single electrode (in three-electrode configuration). Here, the potential varies as a function of time with an applied constant current. Similar to the *CV* technique, the shape of the output curve in this technique provides significant information regarding the charge storage process along with *iR* drop, percentage con-tribution to double-layer and faradic charge storage, and effective series resistance of the supercapacitor device. However, this method has distinguishing results if applied to three-electrode and two-electrode configurations.

As a constant current is maintained throughout the charge–discharge process of the supercapacitor device, unlike the potential step variation in the case of cyclic voltammetry, CCCD technique is more accurate in evaluating the capacitive property.

3.4.4 Electrochemical Impedance Spectroscopy

Electrochemical impedance spectroscopy is a technique in which an AC signal of very small amplitude (to maintain the linearity) is applied to the supercapacitor device to investigate its frequency response. This technique is mostly used for capacitors used in AC line filtering; however, recently it has been found to be effective in characterizing electrochemical capacitors. Through EIS technique, phase angle and complex impedance of the supercapacitor device can be tested with respect to applied frequency range (typically between 0.01 and 1000 Hz). Two major graphical illustrations, i.e., Bode plot and Nyquist plot, are obtained through this technique. While the former establishes a relationship between the phase angle and the frequency, latter one reveals both real and imaginary impedance values of the device. The frequency response recorded by charging–discharging the supercapacitor device is due to the applied potential sweep rate superimposed by a low-amplitude AC signal. The response can therefore be directly correlated to the potential sweep rate applied during cyclic voltammetry technique. Bode plot is quite effective in calculating the gain and phase response of a system under investigation. The shifting in phase at specified frequency values could reveal whether the device is capacitive or not.

EIS technique is particularly useful for calculating the effective series resistance (ESR) value of a supercapacitor device. The ESR value so obtained could help in calculating the specific power of a supercapacitor device in a precise manner. Strangely enough, it has been observed that the impedance spectra of double-layer capacitors, pseudocapacitors, and batteries show almost similar patterns through their behavior over the whole frequency range. The term is synonymous with the solution resistance (R_s) often characterized by the offset value of the EIS spectra from the origin of the impedance axes. This method can provide an accurate measure of the ESR value directly from the obtained impedance spectra.

The Nyquist plot helps in addressing a series of processes occurring in the supercapacitor device. In general, three major portions of the impedance spectra are taken into consideration, i.e., a semicircular region appearing in the high-frequency zone, a linear portion making a slope of 45° with the abscissa called the Warburg element at mid-frequency (knee frequency) zone, and another linear portion with high slope value corresponding to double-layer capacitance appearing at the low-frequency region. There may be an additional portion called the inductive tail (of the order of few micro-Henry, μH) due to inductive effect between the electrodes resulting from non-uniform separation or anomalous magnetism developed on edge atoms. Figure 3.5, respectively, shows typical frequency response in terms of Nyquist and

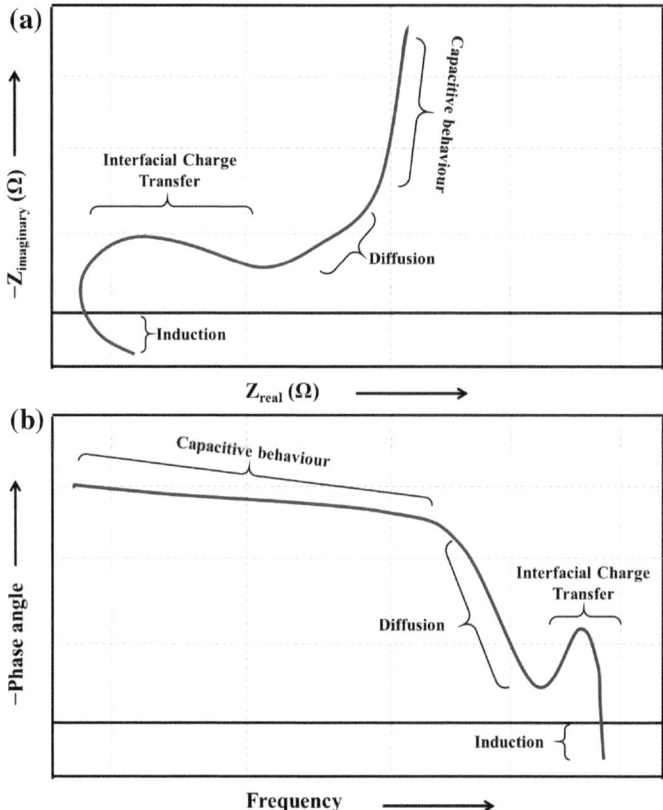

Fig. 3.5 Schematic presentation for the **a** Nyquist and **b** Bode plots for both supercapacitors and batteries

Bode plot. Though similar patterns are often encountered for most of the systems, there may be frequency-dependent processes occurring outside the common frequency range as in the case of solid electrolyte interphase (SEI).

As shown in the Nyquist plot (Fig. 3.5), the semicircular portion of the frequency response represents the charge transfer resistance (often designated as R_{CT}) generated at the electrode/electrolyte interface. The semicircular region is often encountered in the case of supercapacitor devices, though it might get suppressed in the Nyquist plot due to scaling. This phenomenon is also visible in the Bode plot, but not in as a sharp peak which may be due to several limiting factors (Chen et al. 2013; Krishnamoorthy et al. 2017; Shuai et al. 2017; Wei et al. 2015).

The second element, Warburg impedance, is associated with solid-state diffusion processes occurring in the device. Therefore, in typical double-layer capacitors, Warburg impedance generally does not appear in the Nyquist plot due to the absence of concentration gradient in the electrolyte region (Song and Bazant 2013). However, change in the concentration gradient of the bulk electrolyte due to solid-state diffusion at the electrode surface could lead to the appearance of the Warburg element making a sharp 45° slope with the abscissa (*X*-axis).

Like double-layer capacitors, Warburg impedance is the absence in true pseudo-capacitive materials as they rely on only surface redox reactions. In this case, the third type of impedance replaces the Warburg element. Nevertheless, it cannot be concluded that these pseudocapacitive materials are not diffusion controlled. Song and Bazant, in their theoretical model, have explained that these three frequency responses from a device would be clearly distinguishable if the electrode material is composed of homogeneous particles (Song and Bazant 2013). However, it would be difficult to resolute these three regions in case the electrode is composed of particles with high degrees of grain boundaries as in the case of most supercapacitor electrodes (Eftekhari and Mohamedi 2017). Here, the slope of the Warburg impedance element increases enough to merge itself with the third region imparting a notion that no diffusion process is taking place. It should be noted that for an ideal double-layer capacitor, the impedance spectra would show only a straight line coinciding with the positive imaginary impedance axis and intercepting the real impedance axis at the origin (zero point) indicating pure capacitive behavior. This can also be observed in the Bode plot as a gradual increase in the phase angle to reach a value of 90°. Occasionally, a plateau can be observed at a slope of 45° which may be correlated with pseudocapacitive behavior (Chang et al. 2014; Dubal et al. 2017; Wei et al. 2015).

Another factor that affects the Warburg impedance is the ambient temperature. Though supercapacitor performance is hardly affected by nominal thermal fluctuations, the rate capability changes drastically; this can be attributed to the high sensitivity of the diffusion process toward temperature instabilities, as can be seen by a longer Warburg impedance (Kim et al. 2015; Wang et al. 2013). As discussed in the previous section, EIS is mostly used to calculate the ESR value which is essential in deriving the specific power of a supercapacitor device. This can be achieved in both two- and three-electrode configuration. However, as specific power is obsolete for measurements done in a three-electrode configuration, ESR should be evaluated exclusively for two-electrode supercapacitor devices.

3.4.5 *Ideal and Non-ideal Capacitor and the Origin of ESR*

The potential difference (*V*) developed across the two plates of an ideal capacitor is proportional to the stored charge and a generalized mathematical expression for the process can be given as:

$$Q = CV \tag{3.1}$$

where Q is the amount of charge in coulomb (ampere-second), C is the capacitance value in farad, and V is the potential difference. The expressions for both energy and power of the device can also be given by the expressions:

$$E = \frac{1}{2}CV^2 \tag{3.2}$$

$$P = VI \tag{3.3}$$

where E is the energy in joule, P is the power in watt, and I is the current in ampere. As can be inferred from these expressions, the calculations of key parameters like capacitance, energy, and power are quite straightforward for ideal capacitors as there is no energy or power loss and the charge is stored for an indefinite period of time. However, ideal capacitor does not exist in practice and real capacitors suffer from losses and have few critical limitations.

3.4.5.1   Limitation in the Working Potential Window and ESR

Real capacitors can only operate within a specified potential window failing which would result in electrolyte decomposition or device failure. The losses occurring in such a capacitor device are due to resistance at the electrode surface, electrical contacts, and electrolyte resistance. The combination of these resistances would result in the effective series resistance of the device. One of the simplest models can be assumed by combining the capacitor with the ESR in series. The power loss can be expressed as;

$$P_{\text{loss}} = I^2 \times \text{ESR} \tag{3.4}$$

The loss occurs mostly in the form of heat. However, this ESR value can also be crucial in calculating the power of a capacitor:

$$P = \frac{V^2}{4\text{ESR}} \tag{3.5}$$

The heat loss could, under rare circumstances, damage the whole device. So, ESR value is essential in the evaluation of a supercapacitor device.

3.4.5.2   Leakage Current

Unlike ideal capacitors which do not require any additional current to maintain the potential difference across the terminals, real capacitors do rely on currents (though small in magnitude) termed as leakage current to maintain a constant potential difference. Leakage current can be modeled as a resistance connected in parallel to the capacitor in the circuit allowing for gradual discharge of the stored charge (self-discharge). This resistance is synonymous with the term faradic resistance which is inversely related to the applied potential and is governed by a Tafel-type equation (Conway et al. 1997).

Leakage current can be measured through different techniques. One of the methods is to apply a DC potential difference to the device and measure the current that is required to keep the potential difference unchanged in value. Another method is to charge the device to its full potential and measure the potential change over a period of time allowing for self-discharge.

3.4.5.3 Time Constant

Time constant is an important factor to analyze the underlying charge–discharge process of a capacitor. Ideal capacitors backed by electrostatic charge storage would have much smaller time constant parameter than non-ideal capacitors whose charge–discharge process is non-electrostatic (e.g., dielectric adsorption) in nature. The expression for time constant parameter in terms of ESR can be given as:

$$T = \text{ESR} \times C \tag{3.6}$$

Before going into the details of the calculation of key parameters such as net capacitance, working potential window, energy density, and power density, it should be noted that capacitor non-ideality precludes calculation of a true capacitance value for a practical supercapacitor device. Commercial supercapacitors have a specified capacitance value, valid when measured using a specific set of experiments. Other experimental techniques, including CV, EIS, and many long-term potentiostatic and galvanostatic tests, can give very different capacitance values.

3.5 Evaluation of Capacitance

Earlier discussions suggest that the evaluation of capacitance can be done in both two- and three-electrode configurations with the help of the corresponding i-V response during CV and CCCD measurement. Nonetheless, the cyclic voltammetry and CCCD plot in both the cases can have contrasting differences depending on the sample under study. Also, two-electrode configurations are accepted as a representative of what we call a real supercapacitor device, while three-electrode configuration is for only material characterization which can be the first step in checking the feasibility of a material toward charge storage. Both cyclic voltammetry and constant current charge–discharge can also be implemented to optimize the working potential window as illustrated in Fig. 3.6 (Khomenko et al. 2006).

The terms such as specific capacitance, specific energy, and specific power should always be associated with two-terminal devices, while three-electrode configuration should be considered during material characterization only. If "C_T" is the total capacitance of a supercapacitor device, then it can be expressed in terms of the applied potential and state of charge as given below:

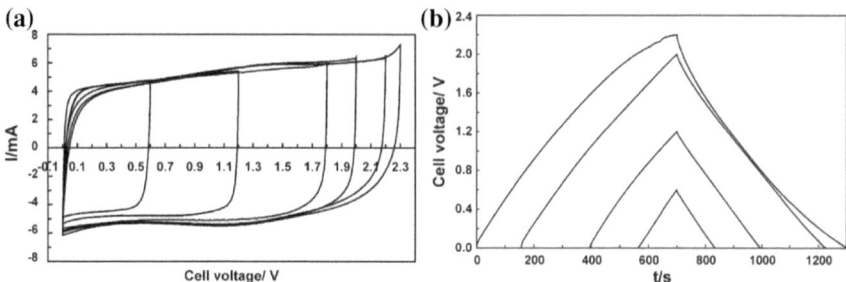

Fig. 3.6 An illustration of V_0 determination methods using **a** CV and **b** CCCD tests (reproduced with permission from Khomenko et al. 2006)

$$C_T = \frac{\Delta Q}{\Delta V} \tag{3.7}$$

Now, the specific capacitance can be calculated from Eq. (3.7) by multiplying the denominator (on the right-hand side of the equation) with a suitable parameter (e.g., mass, volume, area, or length) and the corresponding unit of the specific capacitance is determined. The relevant expression is:

$$C_S = \frac{\Delta Q}{\beta \Delta V} \tag{3.8}$$

where β is the parameter representing mass, volume, area, or length according to which the capacitance, respectively, can be termed as gravimetric, volumetric, areal, or linear capacitance.

3.5.1 From the Cyclic Voltammetry Curve

The calculation for the net specific capacitance can be done from the cyclic voltammetry data by mathematically integrating the absolute area under the curve. The corresponding mathematical expression for the calculation of C_S from a typical cyclic voltammogram is:

$$C_T = \frac{\text{absolute area under the } CV \text{ curve}}{\text{potential sweep rate} \times 2 \times \text{working potential window}} = \frac{\int_0^{2V_0} I(V)\mathrm{d}V}{2 \times r \times V_0} \tag{3.9}$$

where the integral in the numerator is the absolute area under the CV curve, and r is the potential sweep rate.

In the expression given in Eq. (3.9), the factor 2 in the denominator is multiplied to normalize the repeated area of the CV curve consisting of an anodic scan (from

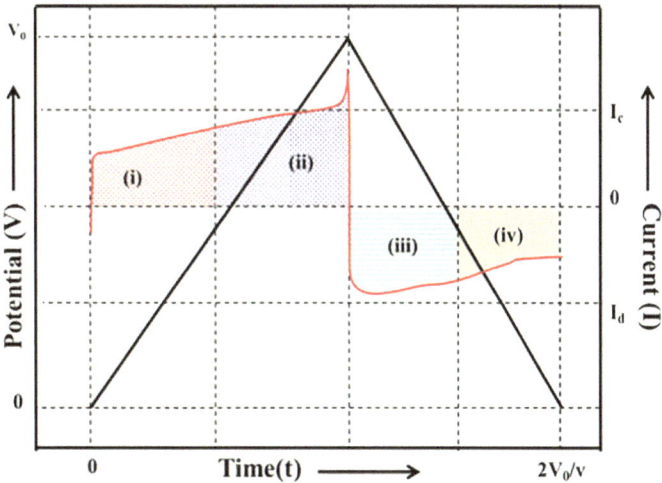

Fig. 3.7 Schematic presentation showing the cyclic voltammogram test result

initial potential to final potential) and again the cathodic scan (from final potential to initial potential) (Zhang and Pan 2015). However, discrepancies have been observed while integrating the different segments (as illustrated in Fig. 3.7).

If mass of the electrode material is taken as the normalizing parameter, then the specific capacitance for a two-terminal supercapacitor device having mass loading of "m" on each electrode surface can be calculated as follows:

$$C_S = \frac{C_T}{2m} \tag{3.10}$$

If C_1 and C_2 are the capacitance of the electrodes in a two-terminal device, then the net device capacitance would be the equivalent of the said capacitances in series; thus, an appropriate expression would be:

$$\frac{1}{C_T} = \frac{1}{C_1} + \frac{1}{C_2} \tag{3.11}$$

Assuming the device to be of symmetric nature, i.e., having same material composition and loading on each electrode, Eq. (3.11) becomes:

$$\frac{1}{C_T} = \frac{2}{C} \Rightarrow C_T = \frac{C}{2} \tag{3.12}$$

(as $C_1 = C_2 = C$, for symmetric two-electrode devices) (3.12)

Now, substituting the value of C_T in Eq. (3.10) with that obtained in Eq. (3.12):

$$C_S = \frac{1}{4}\frac{C}{m} \tag{3.13}$$

The expression for specific capacitance in Eq. (3.13) suggests that for a single electrode (as in the case of a typical three-terminal device, unless specified otherwise), the specific capacitance value can be fourfold higher than that for a two-electrode device constituting the same material and having the same mass loading.

However, the values of specific capacitances from both two- and three-electrode measurement (with same mass loading, area, volume, or length) cannot be compared as the former deals with the device capacitance while the latter is mere material characterization, and therefore could lead to ambiguity. Three-electrode setup is particularly useful in designing supercapacitor devices with two different (asymmetric) electrode materials. If for a given mass loading is taken as a reference, then a pseudocapacitive material could easily exceed a double-layer material in terms of capacitance (or specific capacitance). Therefore, mass loading in each case must be balanced to achieve higher working potential window and to suppress any undesired fluctuations in output current.

The load balancing can be done by taking the help of mass–charge balance equation which is expressed as:

$$m_1 Q_1 = m_2 Q_2 \tag{3.14}$$

where m_1, m_2 and Q_1, Q_2 are mass and state of charge of electrode-1 and electrode-2 in an asymmetric arrangement. The above equation can be rewritten as:

$$m_1 C_1 \Delta V_1 = m_2 C_2 \Delta V_2 \tag{3.15}$$

$$\frac{m_1}{m_2} = \frac{C_2 \Delta V_2}{C_1 \Delta V_1} \tag{3.16}$$

where C_1, C_2 and ΔV_1, ΔV_2 are individual capacitance and optimized working potential window for electrode-1 and electrode-2, respectively, in a three-electrode configuration. These values can be obtained by measuring the capacitance and working potential window for each electrode separately in a three-electrode setup. From Eq. (3.16), the ratio of asymmetric mass can easily be balanced to make unity by simply calculating the capacitance and working potential window in case of each electrode.

3.5.2 From the Constant Current Charge–Discharge Curve

Calculation of capacitance from the constant current charge–discharge curve is simple and straightforward as compared to the calculation techniques for *CV* curves. The plot obtained from the CCCD measurement shows the variation of potential as a function of time under the influence of an applied constant current. Keen observation of the CCCD curve could provide additional information regarding the potential drop (*iR*) due to internal resistance (which is otherwise known as the equivalent

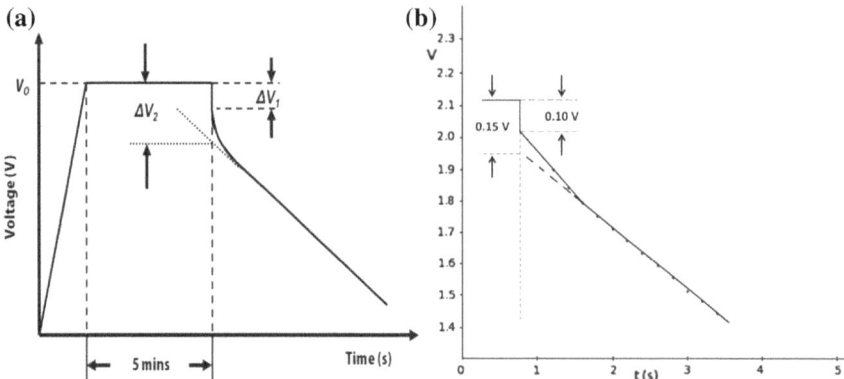

Fig. 3.8 **a** A typical CCCD plot for large SCs with IR drop and steady-state voltage drop marked as $\Delta V1$ and $\Delta V2$, and **b** a real case illustration of the discharge part via Skeleton Tech 1600F SC. (reproduced with permission from Zhang et al. 2015)

series resistance, ESR). The typical iR drop in a commercial supercapacitor from Skeleton Tech has been shown in Fig. 3.8. Sometimes, the accurate method is to extrapolate the discharge curve and calculate the net potential drop (which is in most cases greater than the iR drop calculated in conventional method).

The expression for net capacitance is given as:

$$C_{\mathrm{T}} = \frac{I}{\frac{\Delta V}{\Delta t}} = \frac{I\,\Delta t}{\Delta V} \tag{3.17}$$

Here, I is the constant current applied to the supercapacitor device, and $\Delta V/\Delta t$ is the slope of the discharge curve. During charge–discharge process, iR drop is inevitable due to equivalent series resistance. Therefore, the correct method is to subtract the amount of iR drop from the denominator of Eq. (3.17), and calculate the slope with the help of correct/modified values of time and potential increment.

CCCD technique has another useful application, i.e., to obtain the ESR value by calculating the iR (potential) drop during discharge process. The slope of that particular region of CCCD curve would give the ESR value of the supercapacitor device. This along with the discharge time can be helpful in calculating the value of specific power.

3.5.3   From Electrochemical Impedance Spectroscopy

The impedance spectroscopy reveals both the real and complex impedance components of a supercapacitor device. As capacitive reactance forms the major portion of the imaginary part of the impedance value (inductive effect rarely occurs and that too in the high-frequency region of the AC signal), the capacitance can directly be calculated from below equation:

$$C_{Tf} = \frac{1}{2\pi f \times \text{Im}(Z)} \tag{3.18}$$

where C_{Tf} is the net capacitance value of the device at a particular frequency, f is the frequency of the superimposed AC signal, and Z is the complex impedance value. As most of the commercial supercapacitor devices are unipolar, they are very sensitive toward higher frequencies. Therefore, pseudolinearity should be maintained by keeping the amplitude of the AC signal as small as feasible. Equation (3.18) thus is applicable to low-frequency regions where the capacitance value can be calculated without affecting the device.

According to another report, the frequency-dependent capacitance value can be obtained from the below equations:

$$\text{Re}(C) = \frac{\text{Im}(Z)}{\omega |Z|^2} \tag{3.19}$$

$$\text{Im}(Z) = \frac{\text{Re}(Z)}{\omega |Z|^2} \tag{3.20}$$

where $Z = \sqrt{\text{Re}(Z)^2 + \text{Im}(Z)^2}$ is the complex impedance, $\omega = 2\pi f$ is the angular frequency of the low-amplitude sinusoid signal, $\text{Re}(Z)$ is the real part of the complex impedance, and $\text{Re}(C)$ and $\text{Im}(C)$ are, respectively, the real and imaginary capacitance values. Here, $\text{Im}(C)$ symbolizes the energy dissipation, and hence, the low-frequency $\text{Re}(C)$ only gives the actual capacitance of the device and can be termed as the net capacitance (C_T).

3.6 Evaluation of Specific Energy (E_s)

The specific energy of a supercapacitor device can be calculated by a wide range of methodologies depending upon the parameters and careful selection of the relevant mathematical relations. Specific energy calculation can be done from both cyclic voltammetry and CCCD data.

From cyclic voltammetry, the specific energy can be calculated from the below mathematical equation:

$$E_S^{CV} = \frac{1}{2} C_S (\Delta V)^2 \tag{3.21}$$

where E_s is the specific energy, C_S is the specific capacitance, and ΔV is the potential difference (either for anodic or cathodic scan). The normalization parameter here is still the mass loading of the electrode. The unit of the specific energy obtained via

Eq. (3.21) is in joules which can be converted to watt-hour by dividing the right-hand side expression with a factor of 3600.

From CCCD obtained from EDLC and pseudocapacitors, the plot shows almost a linear charge–discharge curve forming a symmetric triangle. As constant current is being applied during the measurement, simply integrating the area under the triangle would give the capacitance value. In simple terms, the equation would be of the form:

$$E_S^{CCCD} = \int_0^Q V_0 dq = \frac{1}{2} V_0 Q = \frac{1}{2} V_0 I t_c \qquad (3.22)$$

where Q is the net charge acquired by the device with the application of a constant current I over a charging time period of t_c. V_0 is the peak potential value, and only by obtaining the charging time (in this case, discharge time t_d can also be taken as the CCCD curve is symmetric), specific energy can be calculated easily. Dividing Eq. (3.22) with appropriate normalization parameter and a factor of 3600 would produce the desired result in watt-hour scale.

For CCCD plots having nonlinear charge–discharge curves (as in the case of hybrid systems), Eq. (3.22) cannot produce correct results. As the current varies nonlinearly in this case, the integration is to be done incrementally, and thus, it has no simple solutions.

3.7 Evaluation of Specific Power (P_s)

The specific power of a supercapacitor device can also be calculated from the cyclic voltammetry, CCCD, and/or EIS. From CV data, the specific power can be obtained via the following equation:

$$P_S = \frac{1}{2} C_S (\Delta V) r = \frac{1}{2} \left(\frac{Q}{m V_0} \right) V_0 \left(\frac{V_0}{s} \right) = \frac{1}{2} V_0 I \qquad (3.23)$$

Equation (3.23) represents a highly generalized expression for obtaining specific power from the cyclic voltammetry data. Each potential sweep rate (r) would give a single value of specific power depending on the value of the specific capacitance calculated for the same sweep rate. Here, the value of V_0 remains the same.

From CCCD, specific power can be calculated simply dividing Eq. (3.22) by the charging–discharging time. However, these curves can both symmetric and linear (as illustrated in Fig. 3.9a) or highly nonlinear (for hybrid devices as illustrated in Fig. 3.8b).

The required expression in mathematical form for symmetric CCCD curves is:

$$P_S = \frac{E_S^{CCCD}}{t_d} \qquad (3.24)$$

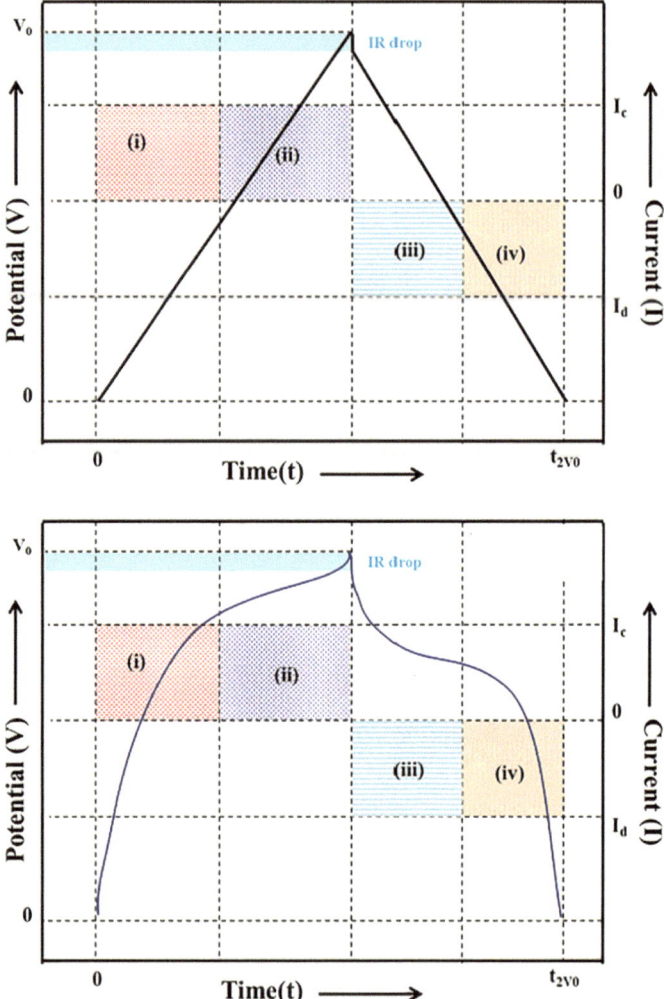

Fig. 3.9 Schematic presentation for the **a** CCCD test results from EDLCs or PCs with linear potential change and **b** from a hybrid supercapacitor with nonlinear potential change over time

where t_d can also be replaced by t_c as the curves are linear and symmetric in nature (i.e., $t_d = t_c$).

The ESR value can be used to calculate the specific power of a supercapacitor device. The ESR value can be obtained from CCCD (*iR* drop) or the impedance spectra. For both the cases, the expression for specific power is the same and given as:

$$P_S = \frac{V_0^2}{4m.\text{ESR}} \qquad (3.25)$$

Combining Eqs. (3.6), (3.21), and (3.21), a relationship between E_S and P_S can be obtained as follows:

$$\frac{E_S}{P_S} = 2\text{ESR}.C_T = 2T \qquad (3.26)$$

Equation (3.26) is of special interest as it establishes a relationship between specific power and energy through the time constant of the supercapacitor device. It suggests that if the capacitance value is increased, then the system response would be slower due to low specific power. Therefore, increasing the working potential window should be given preference in order to keep a good balance between specific energy and power.

Chapter 4
Summary and Focus Point

In this book, different possible electrode fabrication and their arrangement to configure the supercapacitor devices for practical applications have been discussed. The supercapacitors have more than ~100-folds higher power densities and have very long cycle life than traditional batteries but associated with lower values of energy densities impeding their practical application. Therefore, focuses now have to give not only on the development of more active electrode materials and conductive electrolytes, but also onto different possible electrode configurations. Besides the conventional configurations and preparation methodologies, more emphasis on the nonconventional configurations is now given to meet the requirements to use as the energy storage device for the stretchable/flexible SCs as well as for the micro-SCs. Therefore, it is required to develop some intrinsically stretchable/flexible electrode materials and has to find out some noble electrode configurations. In this context, different carbon-based materials like carbon nanotube, nanohorns, graphene, graphene oxide, and their hybrids with metal oxides, sulfides, selenides, etc., are developed and employed as the active electrode materials. Also to achieve better flexibility/stretchability, these active materials are developed onto various substrates, and instead of liquid electrolytes, semisolid electrolytes (gel electrolytes) are developed. Also various measurement methodologies have been established for the accurate evaluation of supercapacitor performances. Before going for the electrochemical measurement, one should have complete knowledge on the chemistry existing among the active (electrode material and electrolyte) and passive (current collector, separator membrane, packaging materials, etc.) components of the proposed SC device. Out of the two main types of measurement systems, the three-electrode analysis mainly employed to evaluate the performance of the developed active electrode material. On the other hand, the two-electrode arrangement strongly recommended for analyzing the complete device performance. Since the specific surface area of carbon materials strongly influence the capacitance values, the efforts are given to synthesize conductive carbon materials with higher porous structure. But due to the mismatch of pore dimension with the size of electrolyte ion, nonlinearity in the charge storage enhancement has been realized. Therefore, the pore size is recommended to optimize according to the size of the electrolyte ions and solvated electrolyte ions for better values of the gravimetric and volumetric capacitances. As discussed in above sections, the specific capacitance, energy/power densities, and cycle life of the SC devices can be calculated by employing the cyclic voltammogram, constant current charge–discharge, and electrochemical impedance spectroscopy. All these parameters are interconnected, so one cannot alter an individual parameter leaving others. Among them, the CCCD (in two-electrode system) can be used for the calculation of all the key parameters, whereas both the CV and CCCD (in three-electrode system) are employed to examine the stable working potential window and specific capacitance of the electrode material.

Here also, we have tried to point out the present practices followed by the researchers for the performance evaluation methods and also discussed the associated inconsistencies. For better comparison, the measurement of all the devices and synthesized active electrode materials should be performed using the same electrochemical instrument with identical electrochemical setups in consistent environment condition. And also, a constant calculation method in every case should be followed.

References

Acerce M, Voiry D, Chhowalla M (2015) Metallic 1T phase MoS_2 nanosheets as supercapacitor electrode materials. Nat Nanotechnol 10:313–318

Bae J, Song MK, Park YJ, Kim JM, Liu M, Wang ZL (2011) Fiber supercapacitors made of nanowire-fiber hybrid structures for wearable/flexible energy storage. Angew Chemie—Int Ed 50:1683–1687

Brousse T, Bélanger D, Long JW (2015) To be or not to be pseudocapacitive? J Electrochem Soc 162:A5185–A5189

Burke A (2007) R&D considerations for the performance and application of electrochemical capacitors. Electrochim Acta 53:1083–1091

Burke A, Miller M (2011) The power capability of ultracapacitors and lithium batteries for electric and hybrid vehicle applications. J Power Sources 196:514–522

Chang Y, Han G, Fu D, Liu F, Li M, Li Y, Liu C (2014) Paper-like N-doped graphene films prepared by hydroxylamine diffusion induced assembly and their ultrahigh-rate capacitive properties. Electrochim Acta 115:461–470

Chen S, He G, Hu H, Jin S, Zhou Y, He Y, He S, Zhao F, Hou H (2013) Elastic carbon foam via direct carbonization of polymer foam for flexible electrodes and organic chemical absorption. Energy Environ Sci 6:2435–2439

Chen W, Rakhi RB, Hu L, Xie X, Cui Y, Alshareef HN (2011) High-performance nanostructured supercapacitors on a sponge. Nano Lett 11:5165–5172

Chen XY, Chen C, Zhang ZJ, Xie DH, Deng X (2013) Nitrogen-doped porous carbon prepared from urea formaldehyde resins by template carbonization method for supercapacitors. Ind Eng Chem Res 52:10181–10188

Choi C, Kim SH, Sim HJ, Lee JA, Choi AY, Kim YT, Lepró X, Spinks GM, Baughman RH, Kim SJ (2015) Stretchable, weavable coiled carbon nanotube/MnO_2/polymer fiber solid-state supercapacitors. Sci Rep 5:9387

Choi D, Blomgren GE, Kumta PN (2006) Fast and reversible surface redox reaction in nanocrystalline vanadium nitride supercapacitors. Adv Mater 18:1178–1182

Conway BE (1991) Transition from "supercapacitor" to "battery" behavior in electrochemical energy storage. J Electrochem Soc 138:1539–1548

Conway BE (1999) Electrochemical supercapacitors, 1st edn. Springer US

Conway BE, Birss V, Wojtowicz J (1997) The role and utilization of pseudocapacitance for energy storage by supercapacitors. J Power Sources 66:1–14

Dong L, Xu C, Li Y, Wu C, Jiang B, Yang Q, Zhou E, Kang F, Yang QH (2016) Simultaneous production of high-performance flexible textile electrodes and fiber electrodes for wearable energy storage. Adv Mater 28:1675–1681

© The Author(s), under exclusive licence to Springer Nature Singapore Pte Ltd. 2018   47
S. Ratha and A. K. Samantara, *Supercapacitor: Instrumentation, Measurement and Performance Evaluation Techniques*, SpringerBriefs in Materials,
https://doi.org/10.1007/978-981-13-3086-5

Dong X, Guo Z, Song Y, Hou M, Wang J, Wang Y, Xia Y, n.d. Flexible and wire-shaped micro-supercapacitor based on Ni(OH)$_2$-nanowire and ordered mesoporous carbon electrodes. Adv Funct Mater 24:3405–3412

Dubal DP, Chodankar NR, Vinu A, Kim D-H, Gomez-Romero P (2017) Asymmetric supercapacitors based on reduced graphene oxide with different polyoxometalates as positive and negative electrodes. ChemSusChem 10:2742–2750

Dupont MF, Donne SW (2015) A step potential electrochemical spectroscopy analysis of electrochemical capacitor electrode performance. Electrochim Acta 167:268–277

Eda G, Fanchini G, Chhowalla M (2008) Large-area ultrathin films of reduced graphene oxide as a transparent and flexible electronic material. Nat Nanotechnol 3:270

Eftekhari A (2018) The mechanism of ultrafast supercapacitors. J Mater Chem A 6:2866–2876

Eftekhari A, Garcia H (2017) The necessity of structural irregularities for the chemical applications of graphene. Mater Today Chem 4:1–16

Eftekhari A, Mohamedi M (2017) Tailoring pseudocapacitive materials from a mechanistic perspective. Mater Today Energy 6:211–229

Electric double layer capacitors for use in hybrid electric vehicles—Test methods for electrical characteristics [WWW Document], 2006 Int Electrotech Comm

Forghani M, Donne SW (2018) Method comparison for deconvoluting capacitive and pseudo-capacitive contributions to electrochemical capacitor electrode behavior. J Electrochem Soc 165:A664–A673

Forghani M, Donne SW (2018) Duty cycle effects on the step potential electrochemical spectroscopy (SPECS) analysis of the aqueous manganese dioxide electrode. J Electrochem Soc 165:A593–A602

Futaba DN, Hata K, Yamada T, Hiraoka T, Hayamizu Y, Kakudate Y, Tanaike O, Hatori H, Yumura M, Iijima S (2006) Shape-engineerable and highly densely packed single-walled carbon nanotubes and their application as super-capacitor electrodes. Nat Mater 5:987–994

Gao Q, Demarconnay L, Raymundo-Piñero E, Béguin F (2012) Exploring the large voltage range of carbon/carbon supercapacitors in aqueous lithium sulfate electrolyte. Energy Environ Sci 5:9611–9617

Ghosh D, Giri S, Mandal M, Das CK (2014) High performance supercapacitor electrode material based on vertically aligned PANI grown on reduced graphene oxide/Ni(OH)$_2$ hybrid composite. RSC Adv 4:26094–26101

Gibson AJ, Donne SW (2017) A step potential electrochemical spectroscopy (SPECS) investigation of anodically electrodeposited thin films of manganese dioxide. J Power Sources 359:520–528

Green R, Staffell I (2016) Electricity in Europe: exiting fossil fuels? Oxford Rev Econ Policy 32:282–303

Guan C, Wang J (2016) Recent development of advanced electrode materials by atomic layer deposition for electrochemical energy storage. Adv Sci 3:1500405

Hao C, Yang B, Wen F, Xiang J, Li L, Wang W, Zeng Z, Xu B, Zhao Z, Liu Z, Tian Y (2016) Flexible all-solid-state supercapacitors based on liquid-exfoliated black-phosphorus nano-flakes. Adv Mater 28:3194–3201

Hu L, Chen W, Xie X, Liu N, Yang Y, Wu H, Yao Y, Pasta M, Alshareef HN, Cui Y (2011) Symmetrical MnO$_2$-carbon nanotube-textile nanostructures for wearable pseudocapacitors with high mass loading. ACS Nano 5:8904–8913

Huang J, Sumpter BG, Meunier V (2008) Theoretical model for nanoporous carbon supercapacitors. Angew Chemie—Int Ed 47:520–524

Huang J, Sumpter BG, Meunier V, Gogotsi YG, Yushin G, Portet C (2010) Curvature effects in carbon nanomaterials: Exohedral versus. J Mater Res 25:1525–1531

Huang L, Chen D, Ding Y, Feng S, Wang ZL, Liu M (2013) Nickel-cobalt hydroxide nanosheets coated on NiCo$_2$O$_4$ nanowires grown on carbon fiber paper for high-performance pseudoca-pacitors. Nano Lett 13:3135–3139

Huang Y, Zhu M, Pei Z, Xue Q, Huang Y, Zhi C (2016) A shape memory supercapacitor and its application in smart energy storage textiles. J Mater Chem A 4:1290–1297

Idoho National Laboratory (2004) FreedomCAR ultracapacitor test manual

IEA (2017) Global energy and CO_2 status report 2017. Glob Energy CO_2 Status Rep, 2017

Jiang H, Ma H, Jin Y, Wang L, Gao F, Lu Q (2016) Hybrid α-Fe_2O_3@Ni(OH)$_2$ nanosheet composite for high-rate-performance supercapacitor electrode. Sci Rep 6:31751

Jussi Pikkarainen (2016) Infographic: fuel savings of up to 10% possible with start-stop systems [WWW Document]. Skelet Technol

Khomenko V, Frackowiak E, Béguin F (2005) Determination of the specific capacitance of conducting polymer/nanotubes composite electrodes using different cell configurations. Electrochim Acta 50:2499–2506

Khomenko V, Raymundo-Piñero E, Béguin F (2006) Optimisation of an asymmetric manganese oxide/activated carbon capacitor working at 2 V in aqueous medium. J Power Sources 153:183–190

Kim B, Chung H, Kim W (2012) All-solid-state flexible supercapacitors based on papers coated with carbon nanotubes and ionic-liquid-based gel electrolytes high-performance supercapacitors based on vertically aligned carbon nanotubes and nonaqueous electrolytes. Nanotechnology 23:289501

Kim SK, Kim HJ, Lee JC, Braun PV, Park HS (2015) Extremely durable, flexible supercapacitors with greatly improved performance at high temperatures. ACS Nano 9:8569–8577

Kim YJ, Horie Y, Ozaki S, Matsuzawa Y, Suezaki H, Kim C, Miyashita N, Endo M (2004) Correlation between the pore and solvated ion size on capacitance uptake of PVDC-based carbons. Carbon N. Y. 42:1491–1500

Kouchachvili L, Yaïci W, Entchev E (2018) Sciencedirect: hybrid battery/supercapacitor energy storage system for the electric vehicles. J Power Sources 374:237–248

Krishnamoorthy K, Pazhamalai P, Sahoo S, Kim S-J (2017) Titanium carbide sheet based high performance wire type solid state supercapacitors. J Mater Chem A 5:5726–5736

Kurzweil P, Chwistek M (2008) Electrochemical stability of organic electrolytes in supercapacitors: spectroscopy and gas analysis of decomposition products. J Power Sources 176:555–567

Lakshmi V, Ranjusha R, Vineeth S, Nair SV, Balakrishnan A (2014) Supercapacitors based on microporous β-Ni(OH)$_2$ nanorods. Colloids Surfaces A Physicochem Eng Asp 457:462–468

Lee YH, Kim Y, Lee TI, Lee I, Shin J, Lee HS, Kim TS, Choi JW (2015) Anomalous stretchable conductivity using an engineered tricot weave. ACS Nano 9:12214–12223

Li N, Chen Z, Ren W, Li F, Cheng H-M, (2012) Flexible graphene-based lithium ion batteries with ultrafast charge and discharge rates. Proc Natl Acad Sci 109:17360 LP–17365

Li P, Kong C, Shang Y, Shi E, Yu Y, Qian W, Wei F, Wei J, Wang K, Zhu H, Cao A, Wu D (2013) Highly deformation-tolerant carbon nanotube sponges as supercapacitor electrodes. Nanoscale 5:8472–8479

Liao Q, Li N, Jin S, Yang G, Wang C (2015) All-solid-state symmetric supercapacitor based on Co_3O_4 nanoparticles on vertically aligned graphene. ACS Nano 9:5310–5317

Ling Z, Ren CE, Zhao M-Q, Yang J, Giammarco JM, Qiu J, Barsoum MW, Gogotsi Y (2014) Flexible and conductive MXene films and nanocomposites with high capacitance. Proc Natl Acad Sci 111:16676–16681

Liu L, Niu Z, Zhang L, Zhou W, Chen X, Xie S (2014) Nanostructured graphene composite papers for highly flexible and foldable supercapacitors. Adv Mater 26:4855–4862

Liu W, Song M-S, Kong B, Cui Y (2017) Flexible and stretchable energy storage: recent advances and future perspectives. Adv Mater 29:1603436

Liu Z, Wu ZS, Yang S, Dong R, Feng X, Müllen K (2016) Ultraflexible in-plane micro-supercapacitors by direct printing of solution-processable electrochemically exfoliated graphene. Adv Mater 28:2217–2222

Liu ZF, Fang S, Moura FA, Ding JN, Jiang N, Di J, Zhang M, Lepró X, Galvão DS, Haines CS, Yuan NY, Yin SG, Lee DW, Wang R, Wang HY, Lv W, Dong C, Zhang RC, Chen MJ, Yin Q, Chong YT, Zhang R, Wang X, Lima MD, Ovalle-Robles R, Qian D, Lu H, Baughman RH (2015) Hierarchically buckled sheath-core fibers for superelastic electronics, sensors, and muscles. Science (80-.) 349:400–404

MarriThese Authors Contributed Equally, SR, Ratha, S, Rout, CS, Behera, JN (2017) 3D cuboidal vanadium diselenide embedded reduced graphene oxide hybrid structures with enhanced supercapacitor properties. Chem Commun 53:228–231

Meher SK, Rao GR (2011) Ultralayered Co_3O_4 for high-performance supercapacitor applications. J Phys Chem C 115:15646–15654

Meng Y, Zhao Y, Hu C, Cheng H, Hu Y, Zhang Z, Shi G, Qu L (2013) All-graphene core-sheath microfibers for all-solid-state, stretchable fibriform supercapacitors and wearable electronic textiles. Adv Mater 25:2326–2331

Merlet C, Péan C, Rotenberg B, Madden PA, Daffos B, Taberna PL, Simon P, Salanne M (2013) Highly confined ions store charge more efficiently in supercapacitors. Nat Commun 4:2701

Miller JM, Miller J, Smith R, Technologies M, n.d. White paper ultracapacitor assisted electric drives for [WWW Document]. Maxwell Technol Inc.

Mombeshora ET, Nyamori VO (2015) A review on the use of carbon nanostructured materials in electrochemical capacitors. Int J Energy Res 39:1955–1980

Naoi K, Simon P (2008) New materials and new confgurations for advanced electrochemical capacitors. Electrochem Soc Interface 17:34–37

Niu Z, Dong H, Zhu B, Li J, Hng HH, Zhou W, Chen X, Xie S (2013) Highly stretchable, integrated supercapacitors based on single-walled carbon nanotube films with continuous reticulate architecture. Adv Mater 25:1058–1064

Pandolfo, AG, Hollenkamp AF (2006) Carbon properties and their role in supercapacitor.pdf. J Power Sources 157:11–12

Parvez K, Wu ZS, Li R, Liu X, Graf R, Feng X, Müllen K (2014) Exfoliation of graphite into graphene in aqueous solutions of inorganic salts. J Am Chem Soc 136:6083–6091

Portet C, Chmiola J, Gogotsi Y, Park S, Lian K (2008) Electrochemical characterizations of carbon nanomaterials by the cavity microelectrode technique. Electrochim Acta 53:7675–7680

Qi D, Liu Z, Liu Y, Leow WR, Zhu B, Yang H, Yu J, Wang W, Wang H, Yin S, Chen X (2015) Suspended wavy graphene microribbons for highly stretchable microsupercapacitors. Adv Mater 27:5559–5566

Qu D (2002) Studies of the activated carbons used in double-layer supercapacitors. J Power Sources 109:403–411

Qu G, Cheng J, Li X, Yuan D, Chen P, Chen X, Wang B, Peng H (2016) A fiber supercapacitor with high energy density based on hollow graphene/conducting polymer fiber electrode. Adv Mater 28:3646–3652

Qu QT, Wang B, Yang LC, Shi Y, Tian S, Wu YP (2008) Study on electrochemical performance of activated carbon in aqueous Li_2SO_4, Na_2SO_4 and K_2SO_4 electrolytes. Electrochem Commun 10:1652–1655

Ratha S, Marri SR, Behera JN, Rout CS (2016) High-energy-density supercapacitors based on patronite/single-walled carbon nanotubes/reduced graphene oxide hybrids. Eur J Inorg Chem 2016:259–265

Ratha S, Samantara AK, Singha KK, Gangan AS, Chakraborty B, Jena BK, Rout CS (2017) Urea-assisted room temperature stabilized metastable β-NiMoO4: experimental and theoretical insights into its unique bifunctional activity toward oxygen evolution and supercapacitor. ACS Appl Mater Interfaces 9:9640–9653

Rudge A, Davey J, Raistrick I, Gottesfeld S, Ferraris JP (1994) Conducting polymers as active materials in electrochemical capacitors. J Power Sources 47:89–107

Samantara AK, Chandra Sahu S, Ghosh A, Jena BK (2015) Sandwiched graphene with nitrogen, sulphur co-doped CQDs: an efficient metal-free material for energy storage and conversion applications. J Mater Chem A 3:16961–16970

Samantara AK, Kamila S, Ghosh A, Jena BK (2018) Highly ordered 1D NiCo$_2$O$_4$ nanorods on graphene: an efficient dual-functional hybrid materials for electrochemical energy conversion and storage applications. Electrochim Acta 263:147–157

Samantara AK, Ratha S (2018) Materials development for active/passive components of a supercapacitor, 1 edn. Springer

Sarkar D, Wang W, Mecklenburg M, Clough AJ, Yeung M, Ren C, Lin Q, Blankemeier L, Niu S, Zhao H, Shi H, Wang H, Cronin SB, Ravichandran J, Luhar M, Kapadia R (2018) Confined liquid-phase growth of crystalline compound semiconductors on any substrate. ACS Nano 12:5158–5167

Schneuwly A, n.d. Distributed ultracapacitor modules to address power and redundancy needs of vehicles [WWW Document]. MAXWELL Technol. White Pap

Shao Y, El-Kady MF, Wang LJ, Zhang Q, Li Y, Wang H, Mousavi MF, Kaner RB (2015) Graphene-based materials for flexible supercapacitors. Chem Soc Rev 44:3639–3665

Shi H, Shi H (1995) Activated carbons and double capacitance. Electrochim Acta 41:1633

Shuai X, Bo Z, Kong J, Yan J, Cen K (2017) Wettability of vertically-oriented graphenes with different intersheet distances. RSC Adv 7:2667–2675

Song J, Bazant MZ (2013) Effects of nanoparticle geometry and size distribution on diffusion impedance of battery electrodes. J Electrochem Soc 160:A15–A24

Sun Y, Sills RB, Hu X, Seh ZW, Xiao X, Xu H, Luo W, Jin H, Xin Y, Li T, Zhang Z, Zhou J, Cai W, Huang Y, Cui Y (2015) A bamboo-inspired nanostructure design for flexible, foldable, and twistable energy storage devices. Nano Lett 15:3899–3906

Tomáš Zedníček (2016) The Tesla Model S, ultracapacitors, and large energy storage [WWW Document]. Eur. Passiv. COMPONENTS Inst.

Tõnisoo A, Kruusma J, Pärna R, Kikas A, Hirsimäki M, Nõmmiste E, Lust E (2013) In situ XPS studies of electrochemically negatively polarized molybdenum carbide derived carbon double layer capacitor electrode. J Electrochem Soc 160:A1084–A1093

Ultracapacitors Still Showing Promise, 2014. 27, 27

Uno M, Tanaka K (2012) Accelerated charge-discharge cycling test and cycle life prediction model for supercapacitors in alternative battery applications. IEEE Trans Ind Electron 59:4704–4712

Vaughan A (2018) Energy storage leap could slash electric car charging times|environment| the guardian. Guard

Wang H, Xu Z, Kohandehghan A, Li Z, Cui K, Tan X, Stephenson TJ, King'Ondu CK, Holt CMB, Olsen BC, Tak JK, Harfield D, Anyia AO, Mitlin D (2013) Interconnected carbon nanosheets derived from hemp for ultrafast supercapacitors with high energy. ACS Nano 7:5131–5141

Wang K, Zhao P, Zhou X, Wu H, Wei Z (2011) Flexible supercapacitors based on cloth-supported electrodes of conducting polymer nanowire array/SWCNT composites. J Mater Chem 21:16373–16378

Wang K, Zou W, Quan B, Yu A, Wu H, Jiang P, Wei Z (2011) An all-solid-state flexible micro-supercapacitor on a chip. Adv Energy Mater 1:1068–1072

Wang W, Liu W, Zeng Y, Han Y, Yu M, Lu X, Tong Y (2015) A novel exfoliation strategy to significantly boost the energy storage capability of commercial carbon cloth. Adv Mater 27:3572–3578

Wei L, Jiang W, Yuan Y, Goh K, Yu D, Wang L, Chen Y (2015) Synthesis of free-standing carbon nanohybrid by directly growing carbon nanotubes on air-sprayed graphene oxide paper and its application in supercapacitor. J Solid State Chem 224:45–51

Wu C, Lu X, Peng L, Xu K, Peng X, Huang J, Yu G, Xie Y (2013) 23-Two-dimensional vanadyl phosphate ultrathin nanosheets for high energy density and flexible pseudocapacitors. Nat Commun 4:2431

Wu X-L, Wen T, Guo H-L, Yang S, Wang X, Xu A-W (2013) Biomass-derived sponge-like carbonaceous hydrogels and aerogels for supercapacitors. ACS Nano 7:3589–3597

Wu Z, Müllen K, Feng X, Parvez K (2013) Graphene-based in-plane micro-supercapacitors with high power and energy densities BT—nature communications. Nat Commun 4:2487

Xiao J, Wan L, Yang S, Xiao F, Wang S (2014) Design hierarchical electrodes with highly conductive $NiCo_2S_4$ nanotube arrays grown on carbon fiber paper for high-performance pseudocapacitors. Nano Lett 14:831–838

Xiao X, Li T, Yang P, Gao Y, Jin H, Ni W, Zhan W, Zhang X, Cao Y, Zhong J, Gong L, Yen WC, Mai W, Chen J, Huo K, Chueh YL, Wang ZL, Zhou J (2012) Fiber-based all-solid-state flexible supercapacitors for self-powered systems. ACS Nano 6:9200–9206

Xiao X, Peng X, Jin H, Li T, Zhang C, Gao B, Hu B, Huo K, Zhou J (2013) Freestanding mesoporous VN/CNT hybrid electrodes for flexible all-solid-state supercapacitors. Adv Mater 25:5091–5097

Xie Y, Liu Y, Zhao Y, Tsang YH, Lau SP, Huang H, Chai Y (2014) Stretchable all-solid-state supercapacitor with wavy shaped polyaniline/graphene electrode. J Mater Chem A 2:9142–9149

Xu J, Wang Q, Wang X, Xiang Q, Liang B, Chen D, Shen G (2013) Flexible asymmetric supercapacitors based upon Co_9S_8 Nanorod//Co_3O_4 @RuO_2 nanosheet arrays on carbon cloth. ACS Nano 7:5453–5462

Yang CM, Kim YJ, Endo M, Kanoh H, Yudasaka M, Iijima S, Kaneko K (2007) Nanowindow-regulated specific capacitance of supercapacitor electrodes of single-wall carbon nanohorns. J Am Chem Soc 129:20–21

Yoo JJ, Balakrishnan K, Huang J, Meunier V, Sumpter BG, Srivastava A, Conway M, Mohana Reddy AL, Yu J, Vajtai R, Ajayan PM (2011) Ultrathin planar graphene supercapacitors. Nano Lett. 11:1423–1427

Yu C, Masarapu C, Rong J, Wei BQM, Jiang H (2009) Stretchable supercapacitors based on buckled single-walled carbon nanotube macrofilms. Adv Mater 21:4793–4797

Yu L, Chen GZ (2016) Redox electrode materials for supercapatteries. J Power Sources 326:604–612

Zang J, Cao C, Feng Y, Liu J, Zhao X (2014) Stretchable and high-performance supercapacitors with crumpled graphene papers. Sci Rep 4:6492

Zhang LL, Zhao X, Stoller MD, Zhu Y, Ji H, Murali S, Wu Y, Perales S, Clevenger B, Ruoff RS (2012) Highly conductive and porous activated reduced graphene oxide films for high-power supercapacitors. Nano Lett 12:1806–1812

Zhang S, Pan N (2015) Supercapacitors performance evaluation. Adv Energy Mater 5:1401401

Zhang Z, Deng J, Li X, Yang Z, He S, Chen X, Guan G, Ren J, Peng H (2015) Superelastic supercapacitors with high performances during stretching. Adv Mater 27:356–362

Zhou C, Zhang Y, Li Y, Liu J (2013) Construction of high-capacitance 3D CoO@polypyrrole nanowire array electrode for aqueous asymmetric supercapacitor. Nano Lett 13:2078–2085

The SpringerBriefs Series in Materials presents highly relevant, concise monographs on a wide range of topics covering fundamental advances and new applications in the field. Areas of interest include topical information on innovative, structural and functional materials and composites as well as fundamental principles, physical properties, materials theory and design. SpringerBriefs present succinct summaries of cutting-edge research and practical applications across a wide spectrum of fields. Featuring compact volumes of 50 to 125 pages, the series covers a range of content from professional to academic. Typical topics might include:

- A timely report of state-of-the-art analytical techniques
- A bridge between new research results, as published in journal articles, and a contextual literature review
- A snapshot of a hot or emerging topic
- An in-depth case study or clinical example
- A presentation of core concepts that students must understand in order to make independent contributions

Briefs are characterized by fast, global electronic dissemination, standard publishing contracts, standardized manuscript preparation and formatting guidelines, and expedited production schedules.

More information about this series at http://www.springer.com/series/10111

SpringerBriefs in Materials